Student Support Materials for AQA

A2 Chemistry

Unit 4: Kinetics, Equilibria and Organic Chemistry

Authors: John Bentham, Colin Chambers, Graham Curtis,
Geoffrey Hallas, Andrew Maczek, David Nicholls.

William Collins's dream of knowledge for all began with the publication of his first book in 1819. A self-educated mill worker, he not only enriched millions of lives, but also founded a flourishing publishing house. Today, staying true to this spirit, Collins books are packed with inspiration, innovation and practical expertise. They place you at the centre of a world of possibility and give you exactly what you need to explore it.

Collins. Freedom to teach.

Published by Collins
An imprint of HarperCollinsPublishers
77-85 Fulham Palace Road
Hammersmith
London
W6 8JB

Browse the complete Collins catalogue at
www.collinseducation.com

© HarperCollinsPublishers Limited 2008

10 9 8 7 6 5 4 3

ISBN-13 978-00-726827-6

John Bentham, Colin Chambers, Graham Curtis, Geoffrey Hallas, Andrew Maczek and David Nicholls assert their moral right to be identified as the authors of this work.

British Library Cataloguing in Publication Data. A Catalogue record for this publication is available from the British Library.

Commissioned by Penny Fowler
Project managed by Alexandra Riley
Edited by Lynn Watkins
Proof read by Patrick Roberts
Design by Newgen Imaging
Cover design by Angela English
Index by Laurence Errington
Production by Arjen Jansen
Printed and bound in Hong Kong by Printing Express

Mixed Sources
Product group from well-managed
forests and other controlled sources
www.fsc.org Cert no. SW-COC-1806
© 1996 Forest Stewardship Council

FSC is a non-profit international organisation established to promote the responsible management of the world's forests. Products carrying the FSC label are independently certified to assure consumers that they come from forests that are managed to meet the social, economic and ecological needs of present and future generations.

Find out more about HarperCollins and the environment at
www.harpercollins.co.uk/green

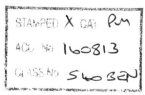

Contents

3.4.1 Kinetics

Essential Notes

The SI units of reaction rate are: $mol\ dm^{-3}\ s^{-1}$.

The SI unit of time is the *second*. The general units for rate are *concentration time^{-1}* so, for some very slow reactions, the rate may be expressed as $mol\ dm^{-3}\ min^{-1}$ or $mol\ dm^{-3}\ hr^{-1}$.

Examiners' Notes

n and m are determined from experimental data.

Examiners' Notes

The value of the rate constant, k, varies with temperature.

Simple rate equations

Rate of a chemical reaction

> **Definition**
>
> The **rate of a chemical reaction** is the change in concentration of a substance in unit time.

This definition is also found in *Collins Student Support Materials: Unit 2 – Chemistry in Action*, section 3.2.2.

The rate depends on the temperature of the reaction (see page 8) and also on the concentrations of the reagents involved. However, the actual relationship at a fixed temperature between the rate of reaction and the reactant concentrations cannot be predicted from the overall chemical equation. Take, for example, a reaction for which the overall chemical equation is:

$$A + 2B \rightarrow C$$

Although the rate of this reaction may well depend on either or both of the reactant concentrations [A] and [B], the rate cannot be assumed to be *directly* proportional (mole per mole) to these concentrations. Instead, the rate is given by the expression:

$$rate \propto [A]^m[B]^n$$

If the rate expression is modified so that the proportional sign is changed into an *equals sign*, it becomes a **rate equation**, to which is added a constant of proportionality, k, called the **rate constant** (or **velocity constant**).

> **Definition**
>
> The **rate equation** expresses the relationship between the rate of reaction and the concentrations of reactants; the constant of proportionality in the rate equation is called the **rate constant**.

At a given temperature, k is constant so that:

$$rate = k[A]^m[B]^n$$

The powers m and n are usually integral, most commonly 0, 1 or 2, and are called the **orders of reaction** with respect to the reactants A and B. The *values* of m and n can never be inferred from the coefficients (numbers of reacting moles) in the stoichiometric equation; the order with respect to a given component is always deduced from experiment.

> **Definition**
>
> The overall **order of a reaction** is the sum of the powers of the concentration terms in the rate equation.

In the above case, $(m + n)$ is the sum of the powers of the concentration terms, so the overall reaction has order $(m + n)$.

Zero-order reactions

Consider a reaction for which the rate equation is:

$rate = k[A]^x$

If x is zero in this equation, then:

$rate = k$

which means that the rate of reaction is *always* constant and independent of the concentration of A, because $[A]^0 = 1$.

If the rate of reaction is constant (Fig 1), a graph of [A] against time is a straight line (Fig 2).

First-order reactions

If x is 1 in the equation above, the rate equation becomes:

$rate = k[A]$

and the reaction is **first order** with respect to A. If [A] doubles, the rate doubles.

Fig 3 shows how concentration varies with time for a first-order reaction. The first-order rate constant, k, has the units s^{-1}, as can be seen by rearranging the rate equation to give:

$k = \dfrac{rate}{[A]}$

in which the units of concentration can be cancelled:

$$\frac{\cancel{(mol\ dm^{-3})}\ s^{-1}}{\cancel{(mol\ dm^{-3})}} = s^{-1}$$

Second-order reactions

The rate equation for a second-order reaction could be:

$rate = k[A]^2$

If so, the reaction is **second order** with respect to A.

Alternatively, the rate equation could be:

$rate = k[A][B]$

If so, the reaction is *first order* with respect to both A and B and the overall order is $(1 + 1) = 2$.

Fig 3 also illustrates how concentration varies with time for a second-order reaction.

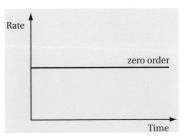

Fig 1
Rate against time for a zero-order reaction

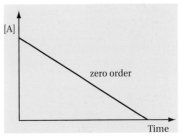

Fig 2
Concentration against time for a zero-order reaction

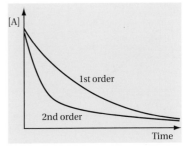

Fig 3
Concentration against time for first- and second-order reactions

A second-order rate constant has units $mol^{-1} dm^3 s^{-1}$. Rearranging the rate equation:

$$k = \frac{rate}{[A][B]}$$

allows units of concentration to cancel:

$$\frac{(mol\,dm^{-3})\,s^{-1}}{(mol\,dm^{-3})(mol\,dm^{-3})} = \frac{s^{-1}}{(mol\,dm^{-3})} = mol^{-1}\,dm^3\,s^{-1}$$

Higher-order reactions

A general form of the rate equation for two reactants A and B is:

$$rate = k[A]^m[B]^n$$

If $n = 1$ and $m = 2$ (or vice-versa) the reaction is **third order** overall.

If $n = 2$ and $m = 2$ (or any other integers that add up to 4) the reaction is **fourth order** overall.

The units for third-order and fourth-order rate constants can be derived using the method shown above. Table 1 on page 7 shows the units of all orders of reaction from order 0 to order 4.

Determination of rate equation

As a reaction proceeds, the *rate* of reaction at fixed temperature decreases because the *concentrations* of the reagents fall as they are being used up. The value of the rate at a particular time can be found by measuring the gradient at that time on a graph of concentration against time (see Fig 4). The rate at the start of the reaction, when the initial concentrations of the reagents are known exactly, is called the **initial rate**.

When the experiment is repeated using different initial concentrations of reagents, the initial rate changes. By recording how different initial concentrations affect the initial rate, chemists can derive the rate equation for a reaction.

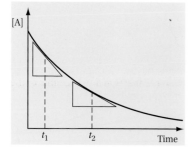

Fig 4
The gradient of the concentration–time curve measures the rate of reaction. The rate is higher at the earlier time (t_1), and falls as the reaction proceeds, eventually falling to zero

Example

The following data were obtained for the reaction:

$$2P + Q \rightarrow R + S$$

Experiment	Initial [P]/mol dm⁻³	Initial [Q]/mol dm⁻³	Initial rate/mol dm⁻³ s⁻¹
1	0.5	0.5	0.002
2	1.0	0.5	0.008
3	1.0	1.0	0.008
4	1.5	1.5	0.018

Determine the order of reaction with respect to components P and Q and deduce the units of the rate constant.

Method

Consider pairs of experiments in which the concentrations of one of the reagents remains constant. This approach establishes the order with respect to the other reagent.

Consider experiments 1 and 2:

[Q] remains constant and [P] is doubled.

The rate increases by a factor of $(0.008/0.002) = 4$ or 2^2 so:

$$rate \propto [P]^2$$

and the reaction is **second order** with respect to P.

Consider experiments 2 and 3:

[Q] is doubled and [P] is kept constant. The rate is unchanged, i.e. the rate is independent of the concentration of Q so:

$$rate \propto [Q]^0$$

and the reaction is **zero order** with respect to Q.

Overall:

$$rate \propto [Q]^0 [P]^2$$

so the overall order is $0 + 2 = 2$

Hence the rate equation is:

$$rate = k[P]^2$$

Comment

The reaction is **second order** overall. Derive the units of this rate constant without referring back to the previous page. The answer is in Table 1 below.

Examiners' Notes

The data given for Experiment 4 are not needed to derive the answer, but can be used as a consistency check. In each of the experiments, the value of k (which is found from the expression $rate/[P]^2$) should be the same. Consistency for Experiments 1 and 4 requires that $0.002/(0.5)^2$ and $0.018/(1.5)^2$ should be the same. Check to see if they are.

There is a very simple pattern to the units of the rate constant and the order of the reaction. This is shown in Table 1 below. To understand how the units of the rate constant can be derived, it is enough to recall that a reaction of order $n + m$ has the rate equation:

$$rate = k \times (concentration)^{n+m}$$

Order of reaction $= n + m$	Rate equation $rate = k \times [reactants]^{n + m}$	Units of the rate constant
0	$rate = k \times concentration^0$	$mol\ dm^{-3}\ s^{-1}$
1	$rate = k \times concentration^1$	s^{-1}
2	$rate = k \times concentration^2$	$mol^{-1}\ dm^3\ s^{-1}$
3	$rate = k \times concentration^3$	$mol^{-2}\ dm^6\ s^{-1}$
4	$rate = k \times concentration^4$	$mol^{-3}\ dm^9\ s^{-1}$

Table 1
Order of reaction and the units of the rate constant

It is by considering kinetic data that the orders of reactions can be deduced. Information of this kind is very important in deciding how to maximise the rate of a reaction, for example in an industrial process.

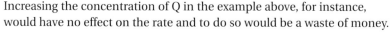
Increasing the concentration of Q in the example above, for instance, would have no effect on the rate and to do so would be a waste of money.

The effect of changes in temperature

An increase in temperature *increases* the *rate* of a reaction. According to kinetic theory, the mean kinetic energy of the particles is directly proportional to the temperature. At higher temperatures, particles have more energy; they move about more quickly, there are more collisions, and these collisions are more energetic. The *increased energy of the collisions* is a much more important factor in affecting the rate than is the relatively slight *increase in the collision rate* when the temperature is raised.

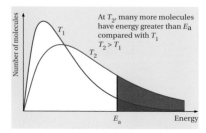

Fig 5

Molecules with energies greater than E_a at different temperatures. The curve for T_2, the higher temperature, is broader and has a lower peak than the curve for T_1. Apart from the origin through which curves at all temperatures pass (there are never any molecules with exactly zero energy), the curve for a higher temperature always lies to the right of that for a lower temperature

Fig 5 shows what happens to the distribution of energies in molecules of a gas when the temperature is increased from T_1 to T_2. For a fixed sample of gas, the total number of molecules is unchanged and the total area under the curve remains constant (*Collins Student Support Materials: Unit 2 – Chemistry in Action*, section 3.2.2). To the right of the maximum, the curve at higher temperature T_2 lies above the one at lower temperature T_1; at the higher temperature there are more molecules with greater energy than there are at the lower temperature.

Particles will react only if, on collision, they have more than a minimum amount of energy known as the **activation energy**.

> **Definition**
>
> The **activation energy** of a reaction is the minimum energy required for the reaction to occur.

Fig 5 shows that, when the activation energy for a reaction is E_a, the number of molecules with energy in excess of E_a is much greater at the higher temperature T_2 than at the lower temperature T_1. The number of collisions between molecules with sufficient energy to react, i.e. the number of productive collisions, and therefore the rate of reaction, will be greater at the higher temperature. Consequently, quite small temperature increases can lead to very large increases in rate (see Fig 6).

This increase in rate arises because the value of the *rate constant, k*, gets bigger when the temperature increases. The *rate equation*:

$$rate = k[A]^m[B]^n$$

has only one possible temperature-dependent feature – the rate constant; the concentrations of reactants do not change with temperature. Thus it is the rate constant k that is affected by the the number of effective collisions in unit time; the number of such collisions is itself dependent on the temperature of the reaction.

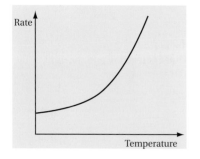

Fig 6

The exponential increase of rate with temperature

The curve in Fig 6, which shows the increase in reaction rate with temperature, also matches the increase in the value of the rate constant as the temperature is increased. The increase in the rate constant with temperature is said to be **exponential**, which means that every time the temperature increases by a certain number of degrees, the rate constant increases by a fixed factor (two-fold, say, or even ten-fold) depending on the size of the temperature increase.

In many chemical reactions that occur reasonably quickly near room temperature (typically biological reactions), a temperature rise of 10 °C increases the rate of reaction by a factor of about two.

Reaction mechanisms and the rate-determining step

Reaction mechanisms

Chemical reactions rarely occur by the simple and straightforward route suggested by the overall stoichiometric equation. Most reactions occur in two or more steps which, when combined, produce the equation for the overall reaction. One of the major tasks of reaction kinetics lies in providing evidence to support or refute the validity of such proposed reaction steps.

A proposed sequence of simple reaction steps is known as a **reaction mechanism**.

> **Definition**
>
> The *reaction mechanism* for a reaction consists of a proposed sequence of discrete chemical reaction steps that can be deduced from the experimentally observed rate equation.

Reactions that occur in steps, and reaction mechanisms, have already been introduced in *Collins Student Support Materials: Unit 2 – Chemistry in Action*, section 3.2.8, where a **free-radical substitution mechanism** is invoked to account for the observed chain-reaction kinetics of the direct chlorination of methane:

Initiation: $\quad Cl_2 \rightarrow 2Cl\bullet$

Propagation: $\quad Cl\bullet + CH_4 \rightarrow \bullet CH_3 + HCl$

$\quad\quad\quad\quad\quad \bullet CH_3 + Cl_2 \rightarrow CH_3Cl + Cl\bullet$

Overall: $\quad CH_4 + Cl_2 \rightarrow CH_3Cl + HCl$

The rate-determining step

In some reactions, the experimental rate equation seems directly related to the process as written in the overall stoichiometric equation. A good example of this is the hydrolysis of 1-bromobutane:

$$CH_3CH_2CH_2CH_2Br + OH^- \rightarrow CH_3CH_2CH_2CH_2OH + Br^-$$

for which the experimental rate equation is:

$$rate = k[CH_3CH_2CH_2CH_2Br][OH^-]$$

The second-order rate equation suggests a single-step nucleophilic substitution mechanism involving direct collision of a hydroxide ion with a molecule of bromobutane; no further explanation is needed.

Examiners' Notes

The *initiation* step involves only a single molecule reacting on its own; this is called a *unimolecular step*.

Both the *propagation steps* involve two species reacting together; these are called *bimolecular steps*.

Termolecular steps (three species reacting simultaneously) are possible, but extremely rare.

Examiners' Notes

Although the rate equation for a reaction cannot be deduced from the stoichiometry of the reaction equation (the rate equation is always derived experimentally), the rate of any step in a mechanism is always proportional to the concentrations of the reactant species in that step, raised to the appropriate powers.

Thus, in the chlorination chain reaction, it is possible to write for the initiation step:

rate = $k_1[Cl_2]$

which is unimolecular, and therefore a first-order step, and for the first propagation step:

rate = $k_2[CH_4][Cl\bullet]$

which is bimolecular, and therefore a second-order step.

However, in the superficially identical hydrolysis of 2-bromo-2-methylpropane:

$$(CH_3)_3CBr + OH^- \rightarrow (CH_3)_3COH + Br^-$$

the experimental rate equation is:

$$rate = k[(CH_3)_3CBr]$$

and the reaction is *first order*, clearly requiring a different *reaction mechanism*.

A proposed reaction mechanism that would fit this rate equation is:

$$(CH_3)_3CBr \xrightarrow{slow} (CH_3)_3C^+ + Br^-$$

$$(CH_3)_3C^+ + OH^- \xrightarrow{fast} (CH_3)_3COH$$

The first, slow step involving the formation of a carbocation determines the overall rate of reaction. It is called the **rate-determining step**.

> ### Definition
> The **rate-determining step** is the slowest step in a multi-step reaction sequence; it dictates the overall rate of reaction.

The rate-determining step above involves only a single parent molecule; this step can be written as:

$$(CH_3)_3CBr \xrightarrow{slow} (CH_3)_3C^+ + Br^- \qquad rate = k_1[(CH_3)_3CBr]$$

and the second step can be written as:

$$(CH_3)_3C^+ + OH^- \xrightarrow{fast} (CH_3)_3COH \qquad rate = k_2[(CH_3)_3C^+][OH^-]$$

However, the kinetics of this step are of no interest, as it can proceed only as fast as the carbocation is formed and then speedily consumed.

The overall reaction is:

$$(CH_3)_3CBr + OH^- \rightarrow (CH_3)_3COH + Br^- \qquad rate = k_1[(CH_3)_3CBr]$$

According to this proposed scheme, the reaction is first order (as found experimentally), so the proposed reaction mechanism is in accord with the observed kinetics.

This is an example of a reaction mechanism where the rate-determining step is the first step in the sequence, so only the reactants in this step can appear in the rate equation. In other cases (mentioned below) the rate-determining step follows after other fast steps; in such cases, the species involved in these fast steps may well appear in the experimental rate equation.

The example below illustrates how kinetic and other data can be used to make proposals about reaction mechanisms and the rate-determining step.

Essential Notes

In a process that involves a sequence of steps, each dependent on the preceding one, the overall rate of conversion is limited by the speed of the slowest step.

Examiners' Notes

It is likely that both mechanisms (the second-order as well as the first-order) are in play in the hydrolysis of both compounds.

For 1-bromobutane, the dissociation into ions will have much higher activation energy than will the approach of a hydroxide ion to the C—Br carbon, which is relatively exposed. So the second-order mechanism will dominate.

However, for 2-bromo-2-methylpropane, the interference of the three quite bulky methyl groups will play a significant role in raising the bimolecular activation energy, as also will their influence in stabilising the tertiary carbocation.

Example

It has been suggested that a reaction of nitrogen dioxide with carbon monoxide can occur in the exhaust gases of motorcars. Kinetic experiments show that the overall reaction is second order in NO_2 and zero order in CO. Suggest a mechanism to account for the observed kinetics and indicate what steps might be taken to support this mechanism.

Method

First write an equation for the overall reaction:

$$NO_2(g) + CO(g) \rightarrow CO_2(g) + NO(g) \qquad rate = k[NO_2]^2$$

Next look for a *rate-determining step* that involves two molecules of NO_2

A possible candidate is:

$$2NO_2 \xrightarrow{\text{slow}} N_2O_4 \qquad \text{known reaction, known product}$$

$$N_2O_4 + CO \xrightarrow{\text{fast}} CO_2 + NO + NO_2 \quad \text{required products, } NO_2 \text{ recycled}$$
$$NO_2(g) + CO(g) \rightarrow CO_2(g) + NO(g) \quad \text{overall, as required}$$

Further investigations

If this is the mechanism, it should be possible to detect dinitrogen tetroxide during the reaction. This could be done spectroscopically. But no N_2O_4 is found at the temperature of the experiment; instead, spectroscopic investigations detect the presence of a short-lived intermediary, NO_3.

Comment

The mechanism above is very unlikely, as no N_2O_4 is found. Also, it takes no account of the presence of NO_3. So, following the scheme above, it is necessary to look for another *rate-determining step* that involves two molecules of NO_2 and also one that produces NO_3.

$$2NO_2 \xrightarrow{\text{slow}} NO_3 + NO \qquad \text{detected product formed}$$

$$NO_3 + CO \xrightarrow{\text{fast}} CO_2(g) + NO_2 \qquad \text{required products, } NO_2 \text{ recycled}$$
$$NO_2(g) + CO(g) \rightarrow CO_2(g) + NO(g) \qquad \text{overall, as required}$$

It cannot be proved that this is the actual mechanism, but it corresponds to all the known facts whereas the earlier one does not.

In cases where the rate-determining step is preceded by other steps in the mechanism, the situation becomes more complicated. An example of this is the oxidation of nitrogen(II) oxide, which exhibits third-order kinetics:

$$2NO(g) + O_2(g) \rightarrow 2NO_2(g) \qquad rate = k[NO]^2[O_2]$$

While it is possible that this reaction occurs by a direct mechanism involving the simultaneous collision of three molecules, this is extremely

Examiners' Notes

An inescapable consequence of the rate-determining step is that the only species that can figure in the final rate equation are those that appear as reactants in steps up to and including the rate-determining step.

Examiners' Notes

This example has been given in order to illustrate how it is that chemists approach problems, and how they try to formulate hypotheses to solve them.

Nothing as demanding as this will be expected in the AQA A2 examination.

Examiners' Notes

This reaction is unusual in that, unlike other chemical reactions, the rate decreases as temperature is increased. This interesting observation lies outside the scope of the specification.

Examiners' Notes

This situation is an example of *How Science Works*.

Knowledge of the oxides of nitrogen lies outside the scope of the AQA A-level specification.

Examiners' Notes

The first two reactions are in equilibrium, with forward (*f*) and backward (*b*) rate constants. Forward and backward rates are equal, so:

$k_{1f}[NO]^2 = k_{1b}[N_2O_2]$

and $[N_2O_2] = \frac{k_{1f}}{k_{1b}}[NO]^2$

Substitution yields:

$$rate = \frac{k_{1f}}{k_{1b}} k_2[NO]^2[O_2]$$

as required by experiment.

unlikely because ternary (three-body) collisions are extremely rare. So a multi-step mechanism needs to be sought.

A possible mechanism that accounts for the known facts involves the formation of a dimer, N_2O_2, as is shown below:

$$2NO \xrightarrow{\text{fast}} N_2O_2 \qquad rate = k_{1f}[NO]^2$$

$$N_2O_2 \xrightarrow{\text{fast}} 2NO \qquad rate = k_{1b}[N_2O_2]$$

$$N_2O_2 + O_2 \xrightarrow{\text{slow}} 2NO_2 \qquad rate = k_2[N_2O_2][O_2]$$

The fact that the overall reaction is first order in oxygen makes it inevitable that the third equation must be the rate-determining step, as it alone involves the concentration of oxygen. Details of this mechanism and how it can lead to an (experimental) rate equation lie well beyond the scope of the specification. However, an outline is given in the Examiners' Notes, shown opposite, in order to illustrate how further investigation can validate a hypothetical suggestion, a good example of *How Science Works.*

3.4.2 Equilibria

Essential Notes

At equilibrium there is no net change in the concentrations of reactants and products, because any tendency for change in one direction is always balanced by an equal and opposite tendency for change in the other direction.

Essential Notes

homogeneous = same phase

In *Collins Student Support Materials: Unit 2 – Chemistry in Action,* section 3.2.3, it was explained that equilibrium is a dynamic process, with forward and reverse reactions proceeding at equal rates, with *no long-term change in the reactant and product concentrations.* It was also explained that the effect of changes in temperature, pressure and concentration on the position of equilibrium could be predicted.

This section introduces a new concept – that of an equilibrium constant – and illustrates how this constant can be used:

- to provide a quantitative measure of the extent of reaction
- to determine the position of equilibrium.

Equilibrium constant K_c for homogeneous systems

Homogeneous system

A *homogeneous system* is one in which all the species present are in the *same phase.* In the case of equilibria, this usually means the liquid phase, although it also includes the possibility of homogeneous reactions in the gas phase.

The equilibrium constant K_c for a system at constant temperature

The **equilibrium constant K_c** is calculated from concentrations at constant temperature of the species involved in the equilibrium.

Definition

The **equilibrium constant** for a reaction is obtained by multiplying together the concentrations of the products, each raised to the power of its coefficient in the stoichiometric equilibrium equation, and dividing this by the concentrations of the reactants, each also raised to the appropriate power.

In general, for a reaction:

$$aA(aq) + bB(aq) \rightleftharpoons cC(aq) + dD(aq)$$

the equilibrium constant is:

$$K_c = \frac{[C]^c[D]^d}{[A]^a[B]^b}$$

where a, b, c and d are the numbers of moles of the species A, B, C and D which appear in the balanced equation for the equilibrium, and [ ] denotes a concentration in mol dm^{-3}.

Units of K_c

The units of K_c depend on the stoichiometry of the chosen equilibrium reaction. For example, the reaction:

$$2A(aq) + B(aq) \rightleftharpoons C(aq)$$

has $K_c = \dfrac{[C]}{[A]^2[B]}$ with units obtained by simplification:

$$\frac{\cancel{mol\,dm^{-3}}}{(mol\,dm^{-3})^2\,\cancel{(mol\,dm^{-3})}} = \frac{1}{(mol\,dm^{-3})^2} = mol^{-2}\,dm^6$$

If there are equal numbers of moles on both sides of the equilibrium equation, then K_c has no units.

Note that the equilibrium constant K_c can also be deduced for a gas-phase reaction. The reaction:

$$2A(g) + B(g) \rightleftharpoons C(g) \quad \text{still has} \quad K_c = \frac{[C]}{[A]^2[B]}$$

with [ ] representing the concentration of gaseous species in mol dm^{-3} of gas phase.

Because the expression for K_c involves the stoichiometric coefficients of the equilibrium equation, the numerical value of K_c, and also its units, are linked uniquely to the equation for which it is defined. Thus, the doubled equation above:

$$4A(g) + 2B(g) \rightleftharpoons 2C(g) \quad \text{has} \quad K_{c1} = \frac{[C]^2}{[A]^4[B]^2} = K_c^2$$

which has a value that is the square of the one for the previous equation.

By the same token, the equilibrium constant for the reverse reaction is the reciprocal of the original equilibrium constant:

$$C(g) \rightleftharpoons 2A(g) + B(g) \quad \text{has} \quad K_{c2} = \frac{[A]^2[B]}{[C]} = \frac{1}{K_c}$$

Calculations using K_c

The calculation of K_c is best illustrated by means of examples, as shown below.

Example

200 g of ethyl ethanoate and 7.0 g of water were refluxed together. At equilibrium, the mixture contained 0.25 mol of ethanoic acid.

Calculate the equilibrium constant for the hydrolysis of ethyl ethanoate.

Method

Determine the concentrations of all the species present using the equilibrium equation below.

Let the *initial* number of moles of ethyl ethanoate be a, that of water b, and the total volume be $V\,dm^3$.

Let x mol each of ester and water react together forming x mol each of alcohol and acid, leaving $(a - x)$ mol of ester and $(b - x)$ mol of water.

Reaction: $CH_3COOC_2H_5(l) + H_2O(l) \rightleftharpoons CH_3COOH(l) + C_2H_5OH(l)$

Initial concn	a/V	b/V	0	0
Equilibrium concn	$(a - x)/V$	$(b - x)/V$	x/V	x/V

Calculation

Initial moles:
ethyl ethanoate: initial number of moles $a = 200/88$ $= 2.27$ mol
water: initial number of moles $b = 7/18$ $= 0.39$ mol

Equilibrium concentrations:
ethanoic acid = ethanol: number of moles (given) x $= 0.25$ mol
ethyl ethanoate: number of moles $(a - x)$ $= (2.27 - 0.25)$ mol
water: number of moles $(b - x)$ $= (0.39 - 0.25)$ mol

Equilibrium concentrations:
$$\begin{aligned}
[CH_3COOC_2H_5] &= (2.27 - 0.25)/V & = 2.02/V \text{ mol dm}^{-3} \\
[H_2O] &= (0.39 - 0.25)/V & = 0.14/V \text{ mol dm}^{-3} \\
[C_2H_5OH] &= [CH_3COOH] & = 0.25/V \text{ mol dm}^{-3}
\end{aligned}$$

Equilibrium constant:

$$K_c = \frac{[C_2H_5OH] \times [CH_3COOH]}{[CH_3COOC_2H_5] \times [H_2O]}$$

Hence $K_c = \dfrac{(0.25/V) \text{ mol dm}^{-3} \times (0.25/V) \text{ mol dm}^{-3}}{(2.02/V) \text{ mol dm}^{-3} \times (0.14/V) \text{ mol dm}^{-3}} = 0.22$ (no units)

Examiners' Notes

$K_c < 1$ so the equilibrium position lies over to the left, i.e. hydrolysis is by no means complete.

Comment

In this case K_c has no units; the concentration units cancel as there are equal numbers of moles on both sides of the equilibrium equation.

Example

2 mol of phosphorus(V) chloride vapour are heated to 500 K in a vessel of volume 20 dm^3. The equilibrium mixture contains 1.2 mol of chlorine. Calculate the value of the equilibrium constant K_c for the decomposition of phosphorus(V) chloride into phosphorus(III) chloride.

Method

Determine the concentrations of the three species present by using the equilibrium equation below.

Let x mol of PCl$_5$ decompose to form x mol each of PCl$_3$ and Cl$_2$.

Reaction:	PCl$_5$(g)	$\rightleftharpoons$	PCl$_3$(g)	+	Cl$_2$(g)
Initial concn	2/V		0		0
Equilibrium concn	(2 − x)/V		x/V		x/V

Calculation

Equilibrium moles:

PCl$_3$ = Cl$_2$:	number of moles (given) x	= 1.2 mol
PCl$_5$:	number of moles (2 − x)	= (2 −1.2) mol

Equilibrium concentrations:

[PCl$_3$] = [Cl$_2$]:	x/V	= 1.2 mol/20 dm^3	= 0.06 mol dm^{-3}
[PCl$_5$]:	(2 − x)/V	= (2 − 1.2) mol/20 dm^3	= 0.04 mol dm^{-3}

Equilibrium constant:

$$K_c = \frac{[PCl_3] \times [Cl_2]}{[PCl_5]}$$

Hence $K_c = \dfrac{0.06 \text{ mol dm}^{-3} \times 0.06 \text{ mol dm}^{-3}}{0.04 \text{ mol dm}^{-3}} = 0.09 \text{ mol dm}^{-3}$

Comment

Because there are two moles of product but only one mole of reactant in the equation then, by cancellation, the units of K_c are mol dm^{-3}.

Qualitative effects of changes in temperature and concentration

The position of equilibrium and the value of the equilibrium constant

In *Collins Student Support Materials: Unit 2 – Chemistry in Action,* section 3.2.3, the effects on the position of equilibrium of the following changes in conditions were considered:

- change in temperature
- change in concentration
- addition of a catalyst.

The effect of changes in reaction conditions can be predicted using **Le Chatelier's principle**, given below:

> **Definition**
>
> *Le Chatelier's principle* states that a system at equilibrium will respond to oppose any change imposed upon it.

Essential Notes

An increase in temperature will always increase the reaction rate and decrease the time required to reach equilibrium.

Change in temperature

A change in temperature changes the value of the equilibrium constant K_c. According to Le Chatelier's principle, the constraint of higher temperature can be relieved if the equilibrium moves in the direction that *absorbs* some of the added heat, thus opposing the change in temperature.

Exothermic reactions

In an **exothermic reaction** heat is given out as the reaction proceeds. This evolution of heat will tend to *raise* the temperature of the reaction mixture. An increase in temperature can be opposed by reaction in the direction which will absorb the added heat and so *decrease* the temperature. Thus, in an exothermic reaction the equilibrium is displaced to the *left* and the equilibrium mixture contains a *lower concentration of products*. The converse is true if the temperature is *decreased*.

Examiners' Notes

For this reaction there are equal numbers of moles on both sides of the equilibrium equation, so the equilibrium constant has no units.

Consider the effect of a change in temperature on the exothermic equilibrium reaction:

$$H_2(g) + I_2(g) \rightleftharpoons 2HI(g) \qquad \Delta H^{\ominus} = -9.6 \text{ kJ mol}^{-1}$$

for which the following values of K_c have been found:

Table 2
Variation of the equilibrium constant with temperature for an exothermic reaction

Temperature/K	Equilibrium constant K_c
298	794
500	160
700	54

K_c *decreases* with *increasing* temperature in an *exothermic* reaction.

Endothermic reactions

In an **endothermic reaction** heat is being taken in as the reaction proceeds. This absorbtion of heat will tend to *lower* the temperature of the reaction mixture. An increase in temperature can be opposed by reaction in the direction which will absorb the added heat so as to *decrease* the temperature. Thus, the equilibrium is displaced to the *right* and the equilibrium mixture contains a *higher concentration of products*. The converse is true if temperature is *decreased*.

Consider the effect of a change in temperature on the endothermic reaction:

$$N_2(g) + O_2(g) \rightleftharpoons 2NO(g) \qquad \Delta H^\ominus = +180 \text{ kJ mol}^{-1}$$

for which the following values of K_c have been found:

Temperature/K	Equilibrium constant K_c
293	4×10^{-31}
700	5×10^{-13}
1500	1×10^{-5}

K_c *increase*s with *increasing* temperature in an *endothermic* reaction.

The effects of changes in temperature on equilibria are summarised below:

ΔH for reaction	Change in temperature	Shift of equilibrium	Yield of product	Equilibrium constant
exothermic	increase	to the left	reduced	reduced
exothermic	decrease	to the right	increased	increased
endothermic	increase	to the right	increased	increased
endothermic	decrease	to the left	reduced	reduced

Change in concentration

At a given temperature the value of the equilibrium constant K_c is fixed. If the concentration of any one species involved in an equilibrium is changed, then the concentrations of all the other species will change so that the value of K_c remains constant.

Consider the ester hydrolysis reaction:

$$CH_3COOC_2H_5(l) + H_2O(l) \rightleftharpoons CH_3COOH(l) + C_2H_5OH(l)$$

for which the equilibrium constant is:

$$K_c = \frac{[CH_3COOH(l)][C_2H_5OH(l)]}{[CH_3COOC_2H_5(l)][H_2O(l)]}$$

If more water is added, the equilibrium position is displaced to the *right* in order to restore the original value of K_c and the equilibrium yields of $CH_3COOH(l)$ and $C_2H_5OH(l)$ are *increased*. In terms of the equilibrium constant, the equilibrium concentrations are affected as follows:

Essential Notes

Increased temperature always shifts the equilibrium in the *endothermic direction*.

Decreased temperature always shifts the equilibrium in the *exothermic direction*.

The value of K_c is altered by changes in temperature.

Table 3
Variation of the equilibrium constant with temperature for an endothermic reaction

Table 4
The effect of temperature on equilibrium

Essential Notes

If the concentration of a reactant is increased (or the concentration of the product is decreased), the equilibrium opposes the change by moving to the right to give more product. The opposite happens if reactant is removed or product is added.

Examiners' Notes

This reaction can be studied in the laboratory. Weighed amounts of $CH_3COOC_2H_5(l)$ and water can be mixed together in a conical flask and an accurately measured volume of concentrated HCl added as a catalyst. The mixture is sealed with a bung and left overnight to reach equilibrium.

Titration with a standard solution of sodium hydroxide gives the total number of moles of acid (both HCl and CH_3COOH) present at equilibrium. A separate titration provides the volume of sodium hydroxide required by the HCl catalyst alone. Subtraction gives the volume of sodium hydroxide required to neutralise the acid (CH_3COOH) that has been produced by hydrolysis. The number of moles of this acid is then calculated and, from this value, the number of moles of all species present at equilibrium is obtained.

an *increase* in $[H_2O(l)]$ is countered by a *decrease* in $[CH_3COOC_2H_5(l)]$ (because some more of the ester now reacts with the added water), and a simultaneous *increase* in both $[CH_3COOH(l)]$ and $[C_2H_5OH(l)]$ (since more of these are formed). Similarly, if *more* $CH_3COOH(l)$ or *more* $C_2H_5OH(l)$ is added, equilibrium yields of $CH_3COOC_2H_5(l)$ and $H_2O(l)$ are *increased*.

If the reactants or products are gases, a change in the pressure of any gaseous species is equivalent to a change in the concentration of that species.

The example below illustrates the application of these principles to a real situation.

Example

Soda water is made by dissolving carbon dioxide in water. Suggest optimum conditions for the manufacture of soda water.

Method

The reaction involved is:

$$CO_2(g) \xrightleftharpoons{\text{water}} CO_2(aq) \qquad \Delta H^{\circ} \text{ is negative}$$

Answer

Soda water goes *flat* if warmed; the process is *exothermic* (heating causes it to go in the endothermic direction, forming more gas). So, cooling will be beneficial to making soda water. Increasing the pressure increases the concentration of the CO_2 gas, hence high pressures are helpful in making soda water.

Optimum conditions

Low temperature and high pressure, which agrees with common sense.

The effect of a catalyst on the equilibrium position and the equilibrium constant

> **Definition**
>
> A **catalyst** *alters the rate of a chemical reaction without itself being consumed.*

A catalyst does not have any effect on the position of equilibrium in a chemical reaction; hence it does not affect the value of the equilibrium constant. Thus a catalyst can *never* affect the yield of chemical processes.

All that a catalyst can do is to *speed up the attainment of equilibrium*; it does so by providing an *alternative path* for the reaction with a *lower activation energy*. (See *Collins Student Support Materials: Unit 2 – Chemistry in Action*, section 3.2.2).

3.4.3 Acids and bases

Brønsted–Lowry acid–base equilibria in aqueous solution

The Brønsted and Lowry definition of acids and bases states that:

> **Definition**
> A **Brønsted–Lowry acid** is a substance which **donates protons** in a reaction; it is a **proton donor**.
> A **Brønsted–Lowry base** is a substance which **accepts protons** in a reaction; it is a **proton acceptor**.

For example, consider the reaction between sodium carbonate and hydrochloric acid:

$$Na_2CO_3 + 2HCl \rightarrow 2NaCl + H_2O + CO_2$$

In ionic form, this can be written

$$CO_3^{2-}(aq) + 2H^+(aq) \rightarrow H_2O(l) + CO_2(g)$$

which shows that the HCl is a proton donor, i.e. a Brønsted–Lowry acid, and that the carbonate ion is a proton acceptor, i.e. a Brønsted–Lowry base.

Consider also the reaction that occurs when concentrated sulfuric acid and concentrated nitric acid are mixed:

$$H_2SO_4 + HNO_3 \rightleftharpoons H_2NO_3^+ + HSO_4^-$$

$$acid\,1 \quad base\,1 \quad\ acid\,2 \quad\ base\,2$$

In this reaction, H_2SO_4 is behaving as a Brønsted–Lowry acid because it is donating a proton to HNO_3 which, rather surprisingly, is behaving as a Brønsted–Lowry base because it is accepting a proton. In the reverse direction, $H_2NO_3^+$ acts as the Brønsted–Lowry acid and HSO_4^- as the Brønsted–Lowry base.

Examiners' Notes

It is perhaps surprising to find HNO_3 and HSO_4^- assuming the role of base, but that is exactly the role they adopt in this equilibrium.

The reaction shown above is an example of an **acid–base equilibrium**, which involves the transfer of a proton from H_2SO_4 to HNO_3. Addition of more H_2SO_4 or HNO_3 will shift the equilibrium to the right, producing more HSO_4^- and $H_2NO_3^+$, whereas addition of HSO_4^- or $H_2NO_3^+$ will move the equilibrium to the left.

Reactions of acids and bases in aqueous solution according to the Brønsted–Lowry scheme almost always involve the transfer of protons under equilibrium conditions:

$$H^+(aq) + B(aq) \rightleftharpoons BH^+(aq)$$

Here, the base B in aqueous solution accepts a proton (i.e. has a proton transferred to it). This happens, for example, when ammonia dissolves in water:

$$NH_3(g) + H_2O(l) \rightleftharpoons NH_4^+(aq) + OH^-(aq)$$

The acid (H_2O) transfers a proton to the base (NH_3) in the forward direction, and the acid (NH_4^+) transfers a proton to the base (OH^-) in the reverse direction.

The most fundamental proton-transfer equilibrium in water involves one water molecule, acting as a Brønsted–Lowry acid, donating a proton to another water molecule, acting as a Brønsted–Lowry base:

$$H_2O(l) + H_2O(l) \rightleftharpoons H_3O^+(aq) + OH^-(aq)$$

acid 1   *base* 1   *acid* 2   *base* 2

Definition and determination of pH

The concentration of hydrogen ions in aqueous solution can vary over such a large range that it is convenient to express it on a logarithmic scale, called the pH scale.

pH is defined as:

> **Definition**
>
> $pH = -log_{10}[H^+(aq)]$

Hydrogen ions are usually represented as $H^+(aq)$, or even simply as H^+ (especially in concentration terms such as $[H^+]$), rather than as the hydrated ion $H_3O^+(aq)$.

Measurement of pH

An indication of the pH of a solution can be obtained using a range of **indicators** (see page 34) whose colour change with pH is known. The pH can conveniently be measured electronically using a pH meter which, for precise work, must first be calibrated against **buffer solutions** (see page 37) whose pH is accurately known (usually at least two reference points are needed).

Strong acids and strong bases

In water, strong acids and strong bases are virtually completely dissociated into ions. At normal concentrations, acid or base molecules are *fully ionised*. There is only a limited number of strong acids in water. The most important of these are the *hydrohalic acids* (HCl, HBr and HI, but not HF which is a weak acid), nitric acid (HNO_3), sulfuric acid (H_2SO_4) and perchloric acid ($HClO_4$).

The vast majority of acids are weak (see page 24) and, at normal concentrations in water, are only partially ionised. The same is true of most bases. In water, the only strong base normally encountered is the hydroxide ion (OH^-).

For strong acids in aqueous solution, the position of the proton-transfer equilibrium lies heavily to the right; proton transfer is virtually complete:

$$HCl(aq) + H_2O(l) \rightleftharpoons H_3O^+(aq) + Cl^-(aq)$$

Essential Notes

The square brackets around the symbol $H^+(aq)$ mean that the pH scale depends on hydrogen ion concentrations expressed in mol dm^{-3}.

Because of the logarithmic scale, pH values are normally quoted to 2 decimal places.

Examiners' Notes

When present in high concentrations, even strong acids are not fully ionised due to interactions between the different species present in solution.

Essential Notes

It would be wrong to suggest that a strong acid must be 100% ionised. Any degree of ionisation appreciably above 50% entitles the acid to be classed as strong.

HCl is a Brønsted–Lowry acid; it donates a proton to water, and is a strong acid because the reaction goes practically to completion. Similarly for the strong acid sulfuric acid:

$$H_2SO_4(aq) + H_2O(l) \rightleftharpoons H_3O^+(aq) + HSO_4^-(aq)$$

In this case, the HSO_4^- ion is behaving as a base in the backward reaction, but it can also behave as an acid:

$$HSO_4^-(aq) + H_2O(l) \rightleftharpoons H_3O^+(aq) + SO_4^{2-}$$

Alkali metal hydroxides are strong bases; they dissociate fully when dissolved in a large volume of water and form aqueous hydroxide ions, OH^- (aq).

$$KOH(s) \xrightarrow{\text{water}} K^+(aq) + OH^-(aq)$$

The hydroxide ion is a Brønsted–Lowry base; it is a proton acceptor. Hydroxide ions react virtually completely with hydrogen ions to form water (*but* see page 22).

$$OH^-(aq) + H_3O^+(aq) \rightleftharpoons 2H_2O(l)$$

which can be simplified to:

$$OH^-(aq) + H^+(aq) \rightleftharpoons H_2O(l)$$

Calculating the pH of an aqueous solution of a strong acid of known concentration

Strong acids can be regarded as fully ionised in dilute aqueous solution. Hence, the hydrogen ion concentration in a dilute solution of a monoprotic acid will be equal to the overall concentration of the acid in that solution.

Example

Calculate the pH of a 0.25 mol dm^{-3} solution of hydrochloric acid.

Method

Since the acid is fully ionised,
$[H^+]$ = overall concentration of HCl = 0.25 mol dm^{-3}

Answer

$$pH = -\log_{10}[H^+]$$

Hence pH = $-\log_{10} 0.25 = 0.60$

Calculating the concentration of an aqueous solution of a strong acid from its pH

Such a calculation is the reverse of the one given above.

Examiners' Notes

Conductivity experiments can be used to differentiate between strong and weak acids or between strong and weak bases. With a simple circuit consisting of carbon electrodes, a bulb and a power supply, the brightness of the bulb indicates the extent of ionisation.

Essential Notes

A **monoprotic strong** acid (e.g. HCl) dissociates in water to produce one mole of protons per mole of acid. H_2SO_4 is a **diprotic** acid.

Examiners' Notes

If the concentration of the acid is greater than 1 mol dm^{-3}, the pH is negative.

At a concentration of 1 mol dm^{-3}, the pH of a strong acid HX is 0.

At a concentration of 2.5 mol dm^{-3}, the pH of this acid is -0.40.

Examiners' Notes

pH can never have very large negative values since these would demand very concentrated acid solutions (a pH of -1 implies a 10 mol dm^{-3} solution of H^+). Very concentrated solutions are impossible to obtain, because the concentration of H^+ is ultimately limited by the solubility of the acid in water as well as by ion association in such solutions.

Example

An aqueous solution of a strong monoprotic acid, HX, has a pH of 2.50. Calculate the concentration of this acid.

Method

$$[H^+] = 10^{-pH}$$

So, convert pH into $[H^+]$ using inverse logarithms (antilogs).

Answer

$$pH = 2.50 = -\log_{10}[H^+]$$

Hence $[H^+]$ = antilog $(-2.50) = 10^{-2.5}$
= 3.2×10^{-3} mol dm^{-3}

Comment

Since the acid is a monoprotic strong acid, fully dissociated, the concentration of HX = $[H^+] = 3.2 \times 10^{-3}$ mol dm^{-3}.

The ionic product of water, K_w

Water can act both as a Brønsted–Lowry acid (donating a proton) and as a Brønsted–Lowry base (accepting a proton). As a result, both hydrogen ions and hydroxide ions exist simultaneously in water according to the equilibrium:

$$H_2O(l) \rightleftharpoons H^+(aq) + OH^-(aq)$$

and, because it is an equilibrium reaction, an equilibrium constant can be derived:

$$K_c = \frac{[H^+(aq)][OH^-(aq)]}{[H_2O(l)]}$$

or, more simply: $K_c = \dfrac{[H^+][OH^-]}{[H_2O]}$ with units mol dm^{-3}.

However, water is only weakly dissociated, so the equilibrium position lies very far over to the left (there are very few hydroxide or hydrogen ions present). Consequently, the concentration of water $[H_2O(l)]$ can be taken to be constant and its value incorporated into K_c. The new constant resulting from this is called the **ionic product of water** and is given the symbol K_w. It is defined as follows:

> **Definition**
>
> $K_w = [H^+][OH^-]$

Examiners' Notes

The ionic constant of water varies with temperature as shown below:

T/K	K_w/mol^2 dm^{-6}
273	0.1×10^{-14}
293	0.7×10^{-14}
298	1.0×10^{-14}
303	1.5×10^{-14}
333	5.6×10^{-14}
373	51.3×10^{-14}

Examiners' Notes

The pH of an iced drink is about 7.5 (see values of K_w above) and that of a steaming mug of tea is around 6.1, yet both are **neutral**.

and has the units (mol dm^{-3})^2, or mol^2 dm^{-6}.

The value of K_w varies with temperature as shown above. K_w increases as temperature increases because the ionic dissociation of water, which involves the breaking of covalent bonds, requires an input of energy; it is an endothermic process.

In **neutral solution** $[H^+] = [OH^-]$ always, so that $K_w = [H^+]^2$ and $[H^+] = \sqrt{K_w}$; at 298 K (when $K_w = 1.0 \times 10^{-14}$), $\sqrt{K_w} = 10^{-7}$, so $pH = -\log_{10}10^{-7} = 7.00$

Because K_w is temperature dependent and $[H^+] = [OH^-]$ *always* in neutral solutions, the pH of a neutral solution must vary with temperature. It is only at the single temperature 298 K, that the pH of pure water has, uniquely, the value of 7.

Using the value of K_w at 298 K, it can be seen that the pH scale from 0 to 14 spans a range of solutions from 1.0 mol dm^{-3} monoprotic strong acid (which has a pH of 0.00) to 1.0 mol dm^{-3} monoacidic strong base (which has a pH of 14.00).

Essential Notes

Strong acids react with strong bases to form salts (like NaCl or K_2SO_4), whose aqueous solutions are *always neutral*; at 298 K, salt solutions have a pH of about 7. Only solutions of salts of weak acids or weak bases (see page 26) have a pH that differs from 7 at 298 K.

Calculating the pH of an aqueous solution of a strong base

Strong bases, like strong acids, are virtually completely ionised in water. So, using the value of K_w, we can calculate the pH of an aqueous solution of a strong base from its molar concentration in mol dm^{-3}.

Example

Calculate the pH at 298 K of a 0.15 mol dm^{-3} solution of sodium hydroxide. At 298 K, $K_w = 1.0 \times 10^{-14}$ mol^2 dm^{-6}.

Method

Since sodium hydroxide is fully ionised in aqueous solution, $[OH^-]$ = overall concentration of NaOH = 0.15 mol dm^{-3}.

Answer

$K_w = [H^+][OH^-] = 10^{-14}$ mol^2 dm^{-6}

Hence $[H^+] = \frac{K_w}{[OH^-]} = \frac{10^{-14}}{0.15} = 6.67 \times 10^{-14}$ mol dm^{-3}

and $pH = -\log_{10}(6.67 \times 10^{-14}) = 13.20$

Examiners' Notes

At a concentration of 1 mol dm^{-3}, the pH of a strong base MOH is 14.

If the concentration of this base is greater than 1 mol dm^{-3}, the pH is greater than 14.

At a concentration of 2.5 mol dm^{-3}, the pH is 14.4.

Calculating the concentration of an aqueous solution of a strong base from its pH

Such calculations are the reverse of that given above.

Example

An aqueous solution of a strong monoacidic base MOH has a pH of 12.60. Calculate the concentration of this base.

Method

Convert pH into $[H^+]$ using inverse logarithms (antilogs), then use the value of K_w to convert $[H^+]$ into $[OH^-]$.

Examiners' Notes

pH can never have a very large positive value, since this too would demand a very concentrated basic solution (a pH of 15 implies a 10 mol dm^{-3} solution of OH^- ions). Very concentrated solutions are impossible to obtain for the same solubility and association reasons that apply in concentrated acids.

Answer

$pH = 12.60 = -\log_{10}[H^+]$

Hence $[H^+] = $ antilog $(-12.60) = 10^{-12.6} = 2.51 \times 10^{-13}$ mol dm^{-3}

and $[OH^-] = \dfrac{K_w}{[H^+]} = \dfrac{10^{-14}}{2.51 \times 10^{-13}} = 4.0 \times 10^{-2}$ mol dm^{-3}

Comment

Since the base is a monoacidic strong base, fully dissociated, the concentration of MOH $= [OH^-] = 4.0 \times 10^{-2}$ mol dm^{-3}.

Weak acids and bases

Strong acids and **strong bases** ionise in water almost completely; **weak acids** and **weak bases** ionise in water only partially.

Weak acids

The extent to which an acid dissociates in water determines whether it is **weak** or **strong**. The strength of the acid is indicated by the position of the equilibrium established when the acid is dissolved in water. If the equilibrium lies to the right, the acid is a strong acid; if it lies to the left, the acid is weak.

Aqueous ethanoic acid is a commonly-encountered weak acid:

$$CH_3COOH(aq) \rightleftharpoons H^+(aq) + CH_3COO^-(aq)$$

The equilibrium here is well over to the left; the dissociation into ethanoate ions and hydrogen ions is only partial. Hence ethanoic acid is a weak acid.

Weak bases

Weak bases too are only partially ionised in aqueous solution. The strength of the base is indicated by the position of the equilibrium established when the base is dissolved in water. If the equilibrium lies to the right, the base is strong; if it lies to the left, the base is weak.

Aqueous ammonia is probably the most commonly-encountered weak base:

$$NH_3(aq) + H_2O(l) \rightleftharpoons NH_4^+(aq) + OH^-(aq)$$

The equilibrium here is well over to the left; the dissociation into ammonium ions and hydroxide ions is incomplete. Hence, ammonia is a weak base.

K_a for weak acids

Definition of K_a

Consider a weak acid HA which dissociates only partially in aqueous solution:

$$HA(aq) \rightleftharpoons H^+(aq) + A^-(aq)$$

The acid dissociation constant, K_a, of acid HA, is defined as:

> **Definition**
>
> $$K_a = \frac{[H^+(aq)][A^-(aq)]}{[HA(aq)]}$$

The units in the expression for K_a can be cancelled:

$$\frac{\cancel{(\text{mol dm}^{-3})} \times (\text{mol dm}^{-3})}{\cancel{(\text{mol dm}^{-3})}} = \text{mol dm}^{-3}$$

The extent to which an acid dissociates in water is determined by the acid dissociation constant. The larger the value of the acid dissociation constant, the stronger the acid. Table 5 shows a number of acid dissociation constants at 298 K.

In general, acids with K_a much *smaller* than about 1 are classed as **weak acids** and those with K_a much *bigger* than 1 are classed as **strong acids.**

The value of K_a for HCl in Table 5 shows that when hydrogen chloride is dissolved in water, the equilibrium lies *very* far to the right, i.e. HCl is fully ionised in dilute aqueous solution; it is therefore a *very* strong acid.

As hydrofluoric and ethanoic acids have small acid dissociation constants, they are only slightly ionised in water. Their dilute aqueous solutions contain many undissociated acid molecules but few hydrogen ions; they are therefore weak acids. The value of the acid dissociation constant for hydrocyanic acid is extremely small, indicating that it is a *very* weak acid.

Approximate expression for K_a

For many weak acids, the expression for K_a:

$$K_a = \frac{[H^+(aq)][A^-(aq)]}{[HA(aq)]}$$

can be approximated if the extent to which HA dissociates is small. If so, then the concentration of acid in the denominator of the equation for K_a can quite reasonably be replaced by $[HA]_{tot}$ which is the *total original concentration* of HA. The hydrogen ion concentration can also be assumed to arise solely from dissociation of the acid and not at all from the ionisation of water. Thus, for a sufficiently weak acid, it is justifiable to write:

$$K_a \approx \frac{[H^+]^2}{[HA]_{tot}}$$

Acid	K_a/mol dm^{-3}
HCl	1.0×10^7
HNO_3	4.0×10^1
HF	5.6×10^{-4}
CH_3COOH	1.7×10^{-5}
HCN	4.9×10^{-10}

Table 5
Acid dissociation constants

> **Essential Notes**
>
> $[HA]_{tot} = [HA]_{eq} + [A^-]_{eq}$ so that
>
> $[HA]_{eq} = [HA]_{tot} - [A^-]_{eq}$
>
> where *eq* denotes the equilibrium concentration of the species in question.
>
> As the acid is weak, and thus not strongly dissociated, $[A^-]_{eq}$ is small enough to be ignored, so that:
>
> $[HA]_{eq} \approx [HA]_{tot}$ and, in addition:
>
> $[H^+]_{eq} = [A^-]_{eq}$

This approximation can be used only in situations when the weak acid *alone* has been added to water. If the solution is then acidified, or if a base or the anion A^- is added, this approximation is no longer valid.

Calculating pH from K_a for a weak acid

The equation above can be used to find the hydrogen ion concentration, and hence the pH, of a weak acid. The following example illustrates the method.

Acid	pK_a
$HClO_2$	1.81
$CH_2ClCOOH$	2.85
HNO_2	3.37
HF	3.46
HCOOH	3.75
CH_3COOH	4.75
HOCl	7.53
HOBr	8.69
NH_4^+	9.25
HCN	9.31
$CH_3NH_3^+$	10.56

Table 6
pK_a values for selected acids

Example

Calculate the pH of a 0.10 mol dm^{-3} solution of methanoic acid. K_a for methanoic acid is 3.6×10^{-4} mol dm^{-3}.

Method

$$K_a = \frac{[H^+(aq)][A^-(aq)]}{[HA(aq)]}$$

$$\approx \frac{[H^+]^2}{[HA]_{tot}} \quad (\textit{weak-acid approximation})$$

Therefore $[H^+] \approx \sqrt{K_a [HA]_{tot}}$

Answer

$K_a = 3.6 \times 10^{-4}$ mol dm^{-3} and $[HA]_{tot} = 0.10$ mol dm^{-3}

Hence $[H^+] = \sqrt{3.6 \times 10^{-4} \text{ mol } dm^{-3} \times 0.10 \text{ mol } dm^{-3}}$

$\qquad\qquad = \sqrt{3.6 \times 10^{-5} \text{ mol } dm^{-3}}$

Thus $[H^+] = 6.0 \times 10^{-3}$ mol dm^{-3} and pH = 2.22

Comment

Increasing the concentration of a weak acid increases the hydrogen ion concentration and decreases the pH of the solution. Increasing K_a has the same effect. The weaker the acid, the higher the pH.

Definition of pK_a

Just as pH = $-\log_{10}[H^+]$, so a similar quantity pK_a can be defined:

> **Definition**
>
> $pK_a = -log_{10}K_a$

A lower value of pK_a suggests a stronger acid (see Table 6); chlorethanoic acid is stronger than ethanoic acid which, in turn, is stronger than the ammonium ion.

Calculating pH from pK_a

These results can now be used in calculations of pH for solutions of weak acids and bases. Some examples are shown below.

Example

Calculate the pH of a 0.025 mol dm^{-3} solution of nitric(III) (nitrous) acid. Use data given in Table 6 to work out your answer.

Method

$$HNO_2(aq) \rightleftharpoons H^+(aq) + NO_2^-(aq)$$

$$K_a = \frac{[H^+(aq)][NO_2^-(aq)]}{[HNO_2(aq)]} \approx \frac{[H^+]^2}{[HNO_2]_{tot}} \quad (\textit{weak-acid approximation})$$

Hence $[H^+] \approx \sqrt{K_a[HNO_2]_{tot}}$

Answer

$[HNO_2]_{tot}$ $= 0.025$ mol dm^{-2} $pK_a = 3.37$

Hence $K_a = 10^{-3.37} = 4.266 \times 10^{-4}$ mol dm^{-3}

$[H^+] \approx \sqrt{4.266 \times 10^{-4} \times 0.025} = 0.003266$ mol dm^{-3}

Thus   pH $= 2.49$

Example

Calculate the pH of a 0.50 mol dm^{-3} solution of ammonium chloride. Use data given in Table 6 in working out your answer.

Method

$$NH_4^+(aq) \rightleftharpoons H^+(aq) + NH_3(aq)$$

$$K_a = \frac{[H^+(aq)][NH_3(aq)]}{[NH_4^+(aq)]} \approx \frac{[H^+]^2}{[NH_4^+]_{tot}} \quad (\textit{weak-acid approximation})$$

Hence $[H^+] \approx \sqrt{K_a[NH_4^+]_{tot}}$

Answer

$[NH_4^+]_{tot}$ $= 0.50$ mol dm^{-2} $pK_a = 9.25$

Hence $K_a = 10^{-9.25} = 5.623 \times 10^{-10}$ mol dm^{-3}

$[H^+] \approx \sqrt{5.623 \times 10^{-10} \times 0.50} = 1.677 \times 10^{-5}$ mol dm^{-3}

Thus   pH $= 4.78$

Comment

The aqueous ammonium ion, formed when ammonium chloride dissolves in water, is a weak acid. The pH of the resulting solution is on the acid side of neutral.

pH curves, titrations and indicators

In analytical chemistry, the variation of pH during acid–base titrations can be used to determine the **equivalence point,** which corresponds to the mixing together of stoichiometrically equivalent amounts of acid and base. A plot of the pH of the solution being titrated against the volume of

Essential Notes

The **equivalence point** is also commonly called the **stoichiometric point**. Ideally, this will coincide with the **end-point** of the titration, which is a term that relates to colour change in an indicator (see page 35) and should only be used in that context.

solution added is known as a **pH curve.** Some typical pH curves are shown in Figs 7 to 9; they represent the following titrations:

- Fig 7: *strong* base added to *strong* acid and *strong* base added to *weak* acid

- Fig 8: *strong* acid added to *strong* base and *strong* acid added to *weak* base

- Fig 9: *weak* base added to *weak* acid.

The shapes of pH curves in acid–base titrations

Figs 7 and 8 (*and their mirror images*) show the overall shapes predicted for titration curves in which 0.1 mol dm^{-3} solutions of various acids and bases are titrated together. The value of pKa for the weak acid and that for the protonated weak base have been chosen, respectively, as 4.75 (like the weak acid ethanoic acid) and 9.25 (like the ammonium ion, which is derived from the weak base ammonia).

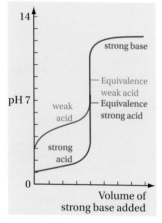

Fig 7
The titration curves of a strong acid (such as HCl) and of a weak acid (such as CH$_3$COOH) with a strong base (such as NaOH)

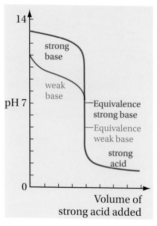

Fig 8
The titration curves of a strong base (such as NaOH) and of a weak base (such as NH$_3$) with a strong acid (such as HCl)

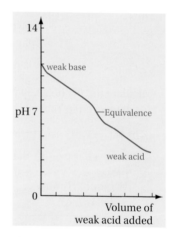

Fig 9
The titration curve of a weak base with a weak acid. The variation of pH with volume near the equivalence point is too gradual to allow for easy detection of equivalence

Examiners' Notes

An aqueous solution of the salt of a *weak acid* and a *strong base* is basic, with pH > 7.

An aqueous solution of the salt of a *strong acid* and a *weak base* is acidic, with pH < 7.

Strong acid–strong base titrations

The two *strong acid–strong base* curves in Figs 7 and 8 are characteristic of the behaviour found for titrations of, for example, hydrochloric acid and sodium hydroxide. The reaction occurring is:

$$HCl(aq) + NaOH(aq) \rightarrow NaCl(aq) + H_2O(l)$$

However, it should be recognised that the species indicated by the short-hand formulae HCl(aq), NaOH(aq) and NaCl(aq) are, in fact, fully dissociated in aqueous solution and exist only as individual H$^+$(aq), Cl$^-$(aq), Na$^+$(aq) and OH$^-$(aq) ions.

If HCl is titrated with NaOH, the initial pH is low (acidic region) and remains low while there is still acid present. Close to the equivalence point, the pH rises rapidly and, at the equivalence point, it reaches the pH of pure water (pH = 7), because the only major ions present [$Na^+(aq)$ and $Cl^-(aq)$] have no effect on pH. After the equivalence point, the excess OH^- ions present cause the pH to rise steeply again to a high value (basic region) where it remains, rising only slowly as more and more base is added.

The curve is traced out in reverse if NaOH is titrated with HCl.

Weak acid–strong base and strong acid–weak base titrations

At the equivalence point in a titration with a strong base (such as NaOH), the pH of a weak acid (such as CH_3COOH) lies above neutrality (pH = 7). At equivalence, the solution contains only the salt of a weak acid. The presence of a Brønsted–Lowry base (the CH_3COO^- ion) means that the pH at equivalence is on the basic side of neutral (pH > 7). A similar argument explains why the pH at equivalence in the case of a weak base–strong acid titration lies on the acidic side of neutrality; the presence of a Brønsted–Lowry acid (NH_4^+ ions) means that the pH at equivalence is less than 7.

Four features are characteristic of titrations involving weak acids or bases with their strong counterparts, as can be seen in Figs 7, 8 and 10:

- **At equivalence**, the pH is not 7, but lies above this for weak acid titrations, or below this for weak base titrations. How to find the equivalence point from an experimental titration curve is shown in Fig 10.

- **Before equivalence**, the pH for weak acid titrations rises less steeply than with strong acid–strong base titrations.

- **After equivalence**, the change in pH follows exactly the trace of the appropriate strong acid–strong base titration. The strong acid or strong base present in excess beyond equivalence totally dominates the shape of the pH curve

- **At the start of the titration**, for a weak acid (Fig 7), the pH rises more steeply than at the start of a strong acid–strong base titration, before flattening out and then rising towards equivalence. The flat portion is due to the formation of a **buffer solution** (see page 37). For a weak base (Fig 8), the pH falls more steeply than at the start of a strong base–strong acid titration, before flattening out and then falling towards equivalence.

Concentrations and volumes of reaction for acids and bases

As the pH curves above have shown, there is an equivalence between a certain volume of an acid of one concentration and another volume of a base of a different concentration. The examples shown below illustrate how to calculate these volumes and concentrations.

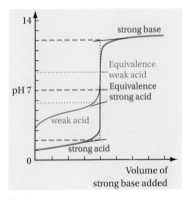

Fig 10
Graphical determination of the equivalence point. The equivalence point lies at the mid-point of the extrapolated vertical portion of the titration curve. The equivalence point for the strong acid is at pH 7.0, mid-way between pH 2.0 and 12.0. For the weak acid, this is at pH 8.9, mid-way between pH 5.8 and 12.0

Example

28.4 cm^3 of a 0.22 mol dm^{-3} solution of Ba(OH)$_2$ are required to neutralise 25.0 cm^3 of a solution of HCl. Calculate the concentration of the HCl solution.

Examiners' Notes

The species HCl, $Ba(OH)_2$ and $BaCl_2$ in aqueous solution are *fully dissociated* into ions and are *not* molecular species. Writing them as if they were molecular, establishes the stoichiometry more easily.

Method

The reaction occurring is:

$$2HCl(aq) + Ba(OH)_2(aq) \rightarrow BaCl_2(aq) + 2H_2O(l)$$

In *Collins Student Support Materials: Unit 1 – Foundation Chemistry*, section 3.1.2, stoichiometric calculations were carried out by determining:

(i)   the number of moles of the known substance

(ii)   the equivalence from the balanced equation, and hence

(iii)   the number of moles of the unknown substance.

Answer

(i)   moles $Ba(OH)_2$: 28.4 cm^3 of a 0.22 mol dm^{-3} solution

$1000 \text{ cm}^3 \equiv 0.22 \text{ mol}$

$28.4 \text{ cm}^3 \equiv \dfrac{0.22 \times 28.4}{1000} = 0.00625 \text{ mol}$

(ii)   stoichiometry: $2 \text{ mol HCl} \equiv 1 \text{ mol } Ba(OH)_2$

(iii)   moles HCl: $2 \times 0.00625 = 0.0125 \text{ mol in } 25.0 \text{ cm}^3$

1000 cm^3 contain $0.0125 \times \dfrac{1000}{25.0} = 0.50 \text{ mol}$

Hence the concentration of the HCl solution is 0.50 mol dm^{-3}.

Comment

It is easy to get the 2:1 ratio the wrong way round. If the base has two hydroxide ions as here then, for roughly equal volumes, they will be matched by close to a doubled concentration of monobasic acid – and vice-versa. It is always worth making this consistency check.

Examiners' Notes

The species NaOH and $Na_2C_2O_4$ are *fully dissociated* into ions in aqueous solution and are *not* molecular species. However, writing them as if they were molecular, establishes the stoichiometry more easily.

Example

25 cm^3 of a solution of ethanedioic acid, $H_2C_2O_4$, is neutralised by 28.6 cm^3 of a 0.28 mol dm^{-3} solution of NaOH. Calculate the concentration of the ethanedioic acid solution.

Method

The reaction occurring is:

$$NaOH(aq) + H_2C_2O_4(aq) \rightarrow Na_2C_2O_4(aq) + H_2O(l)$$

The three steps used in the previous example should be followed.

Answer

(i)   moles NaOH: 28.6 cm^3 of a 0.28 mol dm^{-3} solution

1000 cm$^3 \equiv 0.28$ mol

$$28.6 \text{ cm}^3 \equiv \frac{0.28 \times 28.6}{1000} = 0.0080 \text{ mol}$$

(ii) stoichiometry: 2 mol NaOH $\equiv 1$ mol $H_2C_2O_4$

(iii) moles $H_2C_2O_4$: $\frac{1}{2} \times 0.0080 = 0.0040$ mol in 25 cm^3

$$1000 \text{ cm}^3 \text{ contain } 0.0040 \times \frac{1000}{25} = 0.16 \text{ mol}$$

Hence the concentration of the $H_2C_2O_4$ solution is 0.16 mol dm^{-3}.

Comment

For roughly equal volumes, the dibasic acid will be at approximately half the concentration of the NaOH.

pH calculations for strong acid–strong base titrations

To calculate pH during a titration, use the methods developed above. Two stages are involved in these calculations.

● Find the *number of moles present* of whichever component is *in excess*. If base is being added to acid, the acid will be in excess before equivalence, and the base after equivalence.

● Find the *total volume of solution*. Division of excess moles by total volume gives the *concentration* of the excess component, hence the pH of the solution.

The following examples illustrate this method.

Example

Calculate the pH in the titration of 10.0 cm^3 of a 0.15 mol dm^{-3} solution of HCl at the point when 10.0 cm^3 of a 0.10 mol dm^{-3} solution of NaOH have been added.

Method

Calculate the number of moles of acid originally present, the number of moles of base added, the number of moles of acid in excess, and the total volume of solution.

Answer

moles H^+ originally $= \dfrac{10}{1000} \times 0.15$ $= 0.0015$ mol

moles OH^- added $= \dfrac{10}{1000} \times 0.10$ $= 0.0010$ mol

Examiners' Notes

It is essential to determine the total volume of the solution in order to find the *concentration* of hydrogen ions present. Forgetting the volume is a very easy mistake to make.

moles H^+ in excess	$= 0.0015 - 0.0010$	$= 0.0005$ mol
Total volume	$= 10.0 + 10.0$	$= 20.0$ cm^3
Hence $[H^+]$	$= \dfrac{0.0005 \times 1000}{20}$	$= 0.025$ mol dm^{-3}
pH	$= -\log_{10}[H^+]$	$= 1.60$

Comment

The acid is in excess, so the titration has not yet reached equivalence. Once the excess acid concentration has been determined, the calculation follows the earlier examples shown on pages 21 and 22.

Examiners' Notes

The factor of 2 is used in finding the moles of H^+ added arises because sulfuric acid is a diprotic acid.

Also, it is essential to determine the total volume of the solution in order to find the concentration of hydrogen ions present. Forgetting the volume is a very easy mistake to make.

Example

Calculate the pH in the titration of 16.0 cm^3 of a 0.16 mol dm^{-3} solution of NaOH at the point when 12.0 cm^3 of a 0.10 mol dm^{-3} solution of H_2SO_4 have been added.

Method

The acid:base stoichiometry is 1:2

$$[H^+] = \frac{K_w}{[OH^-]_{xs}}$$

Answer

moles OH^- originally	$= \dfrac{16}{1000} \times 0.16$	$= 0.00256$ mol
moles H^+ added	$= 2 \times \dfrac{12}{1000} \times 0.10$	$= 0.00240$ mol
moles OH^- in excess	$= 0.00256 - 0.00240$	$= 0.00016$ mol
Total volume	$= 16.0 + 12.0$	$= 28.0$ cm^3
Thus $[OH^-]$	$= \dfrac{0.00016 \times 1000}{28}$	$= 0.00571$ mol dm^{-3}
Hence $[H^+]$	$= \dfrac{1.0 \times 10^{-14}}{0.00571}$	$= 1.75 \times 10^{-12}$ mol dm^{-3}
pH	$= -\log_{10}[H^+]$	$= 11.76$

Comment

The base is in excess, so the titration has not yet reached equivalence. Once the excess base concentration has been determined, the calculation follows the same procedure as shown in the examples on pages 23 and 24.

pH calculations for weak acid–strong base titrations

The calculation of pH during these titrations involves using the methods developed earlier. The techniques used depend on how far the titration has progressed.

- **Before equivalence:** the relative proportions of weak acid and its anion present have to be determined and then used in the expression for K_a.

- **After equivalence:** the excess of strong base has to be found, together with the total volume of the solution, and the resulting concentration used to determine pH as in a strong base calculation (see the example on page 23). Each of these techniques is demonstrated in the following examples.

Examiners' Notes

Before equivalence, the mixture of the weak acid and its anion behaves as a **buffer solution** (described later and illustrated in Fig 13).

Example

Calculate the pH in a titration when 10.0 cm^3 of a 0.10 mol dm^{-3} solution of NaOH has been added to 10.0 cm^3 of a 0.25 mol dm^{-3} solution of ethanoic acid (K_a = 1.76 ×10^5 mol dm^{-3}).

Method

$$CH_3COOH(aq) \rightleftharpoons H^+ (aq) + CH_3COO^- (aq)$$

$$K_a = \frac{[H^+][CH_3COO^-]}{[CH_3COOH]}$$

$$\text{So } [H^+] = K_a \times \frac{[CH_3COOH]}{[CH_3COO^-]}$$

Answer

moles CH$_3$COOH originally $= \frac{10.0}{1000} \times 0.25 \qquad = 0.0025$ mol

moles OH$^-$ added $= \frac{10.0}{1000} \times 0.1 \qquad = 0.0010$ mol

moles CH$_3$COO$^-$ formed $\qquad\qquad\qquad\quad = 0.0010$ mol

moles CH$_3$COOH remaining $= 0.0025 - 0.0010 \quad = 0.0015$ mol

so [CH$_3$COOH] $\qquad\qquad = 0.0015/V$

and [CH$_3$COO$^-$] $\qquad\qquad = 0.0010/V$

Since both the ethanoic acid and the ethanoate ions exist together in the same overall volume, *concentration ratio = mole ratio.*

The volumes cancel so $[H^+] = K_a \times$ mole ratio $= 1.76 \times 10^{-5} \times \frac{0.0015}{0.0010}$

$$= 2.64 \times 10^{-5} \text{ mol dm}^{-3}$$

$$pH = -\log_{10}[H^+] \qquad = 4.58$$

Comment

The pH is in the acidic region. Note that it is only the *ratio* of concentrations of weak acid and its anion that matters here, so there is no need to consider the total volume since mole ratio ≡ concentration ratio.

Example

Calculate the pH in a titration when 16.0 cm^3 of a 0.16 mol dm^{-3} solution of NaOH has been added to 12.0 cm^3 of a 0.20 mol dm^{-3} solution of ethanoic acid (K_a = 1.76 × 10^{-5} mol dm^{-3}).

Method

Since the strong base is in excess, the pH will be exactly the same as that in any equivalent strong acid–strong base titration.

Answer

In the example on page 32, 12.0 cm^3 of 0.10 mol dm^{-3} solution of H_2SO_4 were used, which is exactly equivalent to 12.0 cm^3 of a 0.20 mol dm^{-3} solution of ethanoic acid. Furthermore, the final volume in both titrations is the same.

Hence pH = 11.76

Comment

The pH is in the basic region. Note that it is *essential* here to take into account the total volume of the final solution.

The pH of a weak acid at half-equivalence

> **Definition**
>
> At **half-equivalence,** *exactly one-half of the equivalence volume of strong base has been added to the weak acid.*

Examiners' Notes

This **half-equivalence** relationship provides a method for determining pK_a by measuring the pH at half-equivalence.

This titration value has particular significance as it falls at the point where [HA] = [A$^-$] for the weak acid HA. The expression for K_a can be used to find the resulting pH through the equation:

$$[H^+] = K_a \times \frac{[HA]}{[A^-]}$$

The condition that the concentrations in numerator and denominator are the same (half-equivalence) means that:

$$[H^+] = K_a \quad \text{and} \quad pH = pK_a$$

This is shown graphically in Fig 13.

> **Definition**
>
> At **half-equivalence,** *the pH of the solution of a weak acid has the same value as pK_a.*

Examiners' Notes

The additional concentration of H$^+$ ions produced from HIn is too small to affect the final pH of the solution.

Indicators and their range of action

An **acid–base indicator** is a water-soluble, weak organic acid whose acid form (HIn) and base form (In$^-$) have different colours. At least one of the two colours needs to be intense, so that the addition of very little indicator (just a drop or two) will produce a clearly visible colour without appreciably disturbing the acid–base equilibrium to which it has been added.

At the equivalence-point of an acid–base titration, the pH changes very rapidly through several units of pH, and the indicator equilibrium:

$$HIn \rightleftharpoons H^+ + In^-$$

$$colour\ 1 \qquad\qquad\qquad colour\ 2$$

swings from almost *all HIn* to virtually *all In$^-$* (or vice-versa). Thus, if a few drops of indicator have been added, there is an accompanying change from *colour* 1 to *colour* 2, or vice-versa, as the indicator equilibrium moves position under the influence of the changing concentration of H$^+$ ions during the titration. This colour change is used to *indicate* the equivalence point (or, more correctly, the end-point – see below) of the titration.

The dominant species at low pH (when H$^+$ ions are abundant) is the undissociated acid HIn. An abundance of H$^+$ ions pushes the indicator equilibrium to the left. Conversely, at high pH, the anion In$^-$ will dominate. Thus, at low pH the solution has the characteristic colour of HIn and at high pH it has the characteristic colour of In$^-$. An indicator changes colour from only HIn to only In$^-$ over quite a narrow range of pH.

Indicator	Colour 1 acid (HIn)	pH range	Colour 2 base (In$^-$)
thymol blue	red	1.2 to 2.8	yellow
methyl orange	red	3.2 to 4.4	yellow
methyl red	red	4.8 to 6.0	yellow
litmus	red	5.0 to 8.0	blue
bromothymol blue	yellow	6.0 to 7.6	blue
phenol red	yellow	6.6 to 8.0	red
phenolphthalein	colourless	8.2 to 10.0	pink
alizarin yellow	yellow	10.0 to 12.0	red

Table 7
Characteristics of some common indicators

The end-point
The volume of titrant added to give a hydrogen ion concentration such that [HIn] = [In$^-$] is called the **end-point** of the titration. The equivalence volume of titrant can be determined with precision when the indicator chosen has an end-point that coincides with the equivalence point of the titration.

The choice of indicator for a titration
An appropriate indicator for a given titration is best chosen by considering the specific pH curve for that titration. An indicator is appropriate if the rapid change of pH at equivalence overlaps the range of activity of the indicator. The reasons for the choice that is made can be seen by referring to Figs 11 and 12.

Fig 11 shows the pH curves for both a strong acid and a weak acid titrated with a strong base. In the strong acid case, the pH at equivalence is 7; for the weak acid, the pH at equivalence is shown as being 8.8 (as it is when a 0.1 mol dm^{-3} solution of ethanoic acid is titrated with a 0.1 mol dm^{-3} solution of sodium hydroxide).

Examiners' Notes

The end-point (which refers to the *indicator*) should not be confused with the equivalence point (which refers to the *titration*).

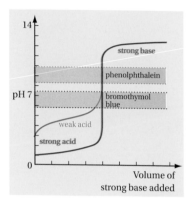

Fig 11
Titration of a strong acid and a weak
acid with a strong base

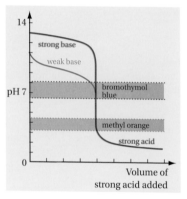

Fig 12
Titration of a strong base and a weak
base with a strong acid

For the strong acid titration, bromothymol blue has a range that includes the equivalence value (pH = 7), so this would be a suitable indicator to choose. However, the change in pH is so steep and extended near equivalence that one drop of base is enough to swing the pH from 3 to 11. Any indicator in this extended range would be as good as bromothymol blue; as a result the choice of indicator is not very critical.

In the weak acid titration, an indicator with a range close to 7 would change colour before equivalence had been reached. With a pH at equivalence close to 9 in this titration, phenolphthalein would be a very suitable indicator.

Fig 12 shows the equivalent situation for a strong base and a weak base titrated with a strong acid. In the case of the strong base, the pH at equivalence is 7 as before; for the weak base, the pH at equivalence is shown as as being 5.2 (as it is when a 0.1 mol dm^{-3} aqueous solution of ammonia is titrated with a 0.1 mol dm^{-3} solution of hydrochloric acid).

Bromothymol blue has a range that includes pH = 7 but, once again, any indicator that changes colour in the pH range from 3 to 11 would be equally suitable. With a pH at equivalence close to 5 in this titration, methyl orange or methyl red would be very suitable indicators.

The pH changes at equivalence of several different titrations are shown in Table 8.

Acid	Base	pH range at equivalence	Choice of indicator (see Table 7)
HCl *strong*	NaOH *strong*	3 to 11	any from methyl orange downwards. The pH change here is over a very wide range
CH$_3$COOH *weak*	NaOH *strong*	7 to 11	any from phenol red downwards
HCl *strong*	NH$_3$ *weak*	3 to 7	methyl orange or methyl red
CH$_3$COOH *weak*	NH$_3$ *weak*	no sharp change	no suitable indicator, nor is the titration suitable with a pH meter since the pH variation shows no abrupt changes
HCl *strong*	Na$_2$CO$_3$ *weak*	2.5 to 5.5 *1st equivalence*	any from methyl orange to methyl red
		6.5 to 9.5 *2nd equivalence*	phenol red, thymol blue or phenolphthalein
H$_2$C$_2$O$_4$ *weak*	NaOH *strong*	1.5 to 3.5 *1st equivalence*	methyl orange or thymol blue; the pH change only just falls in the range of either indicator, so a titration using a pH meter is recommended
		5 to 11 *2nd equivalence*	any from bromothymol blue downwards

Table 8
The choice of an indicator for a
given titration

Buffer action

Definition

A **buffer solution** is one that is able to resist changes in pH when small amounts of acid or base are added. It is also able to maintain its pH in the face of dilution with water.

An **acidic buffer** is one that maintains a solution at a pH below 7 and, typically, consists of a weak acid and one of its salts (to provide the anion, which acts as a base).

A **basic buffer** is one that maintains a solution at a pH above 7 and typically consists of a weak base and one of its salts (to provide the cation, which acts as an acid).

Examiners' Notes

A solution containing ethanoic acid and sodium ethanoate would make an *acidic buffer*.

A solution containing ammonium chloride and ammonia would make a *basic buffer*.

Qualitative explanation of buffer action

Consider an acidic buffer consisting of a solution of ethanoic acid and sodium ethanoate. There is an equilibrium between the components of this solution:

$$CH_3COOH(aq) \rightleftharpoons CH_3COO^-(aq) + H^+(aq)$$

for which the equilibrium constant K_a, the acid dissociation constant, is

$$K_a = \frac{[CH_3COO^-][H^+]}{[CH_3COOH]}$$

which can be rearranged to give

$$[H^+] = K_a \frac{[CH_3COOH]}{[CH_3COO^-]}$$

so that the pH depends on the value of K_a and on the *ratio of the concentrations of the acid and anion (the base)*.

By looking at the equilibrium equation, or at the equilibrium expression for the hydrogen ion concentration, it is clear that:

- pH = pK_a when $[CH_3COOH] = [CH_3COO^-]$, so that the buffer pH will be on the acid side of neutrality.

- Addition of a *small* quantity of hydrogen ions will move the equilibrium to the left, causing ethanoate ions (in plentiful supply) to lower the excess hydrogen ion concentration by forming a little more ethanoic acid.

- Since the ethanoate ion and the ethanoic acid concentrations are both very large, this small increase from ethanoate ions to ethanoic acid

$$A^- + H^+ \rightarrow HA$$

    will not change either the concentration of the acid or of the ion very much.

- As a result, since the ethanoate ion and ethanoic acid concentrations remain much the same, then so clearly does their ratio; the equations above show that the hydrogen ion concentration, and hence the pH, will also remain approximately constant.

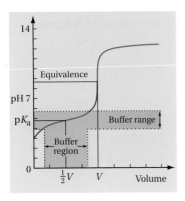

Fig 13

The buffer region for a typical weak acid such as ethanoic acid titrated with a strong base

By direct analogy, the pH remains constant on the addition of small amounts of a base. In this case:

$$HA + OH^- \rightarrow A^- + H_2O$$

and the number of moles of OH^- added gives the number of moles of A^- formed as a result.

From the expression for $[H^+]$ above, it is also clear that dilution will not affect the pH since the *ratio of concentrations* will always be unchanged under dilution.

The presence of appreciable concentrations of both HA and A^- allows $[A^-]$ and $[HA]$ to remain virtually constant for small additions of acid or base and for the ratio of $[HA]$ to $[A^-]$ to remain constant on dilution. Hence $[H^+]$ and pH do not change.

The buffer region and buffer range

The small variation of pH when acid or base is added to a buffer can be understood with reference to the appropriate titration curve. Using as an example a weak acid (Fig 13), it is clear that in the region where the concentrations of acid and anion are similar (close to half-equivalence) the pH remains fairly constant (to within less than ±0.1 pH unit), even when small amounts of strong acid or base are added to the weak acid–anion mixture.

The region over which solutions can show buffer action is called the **buffer region** for the weak acid concerned. The **buffer range** is a term which describes the range of pH values in which a buffer can be prepared using a given weak acid. This range is usually taken to be $pK_a \pm 1$. For ethanoic acid, such a buffer range would be from about pH 3.7 to about pH 5.7.

Calculating the pH of a buffer solution
Acidic buffers

The pH of an acidic buffer solution can be calculated using the weak acid equation:

$$K_a = \frac{[H^+][A^-]}{[HA]}$$

leading to the expression

$$[H^+] = K_a \frac{[\text{weak acid}]}{[\text{salt}]}$$

where [salt] indicates the concentration of the anion of the weak acid present in solution.

A buffer solution may be made either by mixing solutions of the two buffer components (e.g. ethanoic acid and sodium ethanoate) or by partial neutralisation of the weak acid with a strong base (e.g. by adding about half of the stoichiometric amount of sodium hydroxide to a solution of ethanoic acid). In either case, an equimolar mix of components, or a half-neutralised weak acid or weak base, will have:

$$[H^+] = K_a \qquad \text{and hence} \qquad pH = pK_a$$

Two instances are shown in the following examples.

Example

Calculate the pH of a buffer made by mixing 14.0 cm^3 of a 2.0 mol dm^{-3} solution of ethanoic acid ($K_a = 1.74 \times 10^{-5}$ mol dm^{-3}) with 15.0 cm^3 of a solution of 1.50 mol dm^{-3} sodium ethanoate.

Method

Use the expression

$$[H^+] = K_a \frac{[CH_3COOH]}{[CH_3COO^-]}$$

Answer

1000 cm^3 of the acid solution contain 2.0 mol CH$_3$COOH

Thus $\qquad$ 14.0 cm$^3 \equiv \dfrac{2.0 \times 14.0}{1000}$

$\qquad\qquad\qquad = 0.0280$ mol CH$_3$COOH

1000 cm^3 of the salt solution contains 1.50 mol CH$_3$COO$^-$

Thus $\qquad$ 15.0 cm$^3 \equiv \dfrac{1.5 \times 15.0}{1000}$

$\qquad\qquad\qquad = 0.0225$ mol CH$_3$COO$^-$

and $\qquad$ [H$^+$] $\qquad = K_a \times \dfrac{0.0280/V}{0.0225/V}$

$\qquad\qquad\qquad = 1.74 \times 10^{-5} \times 1.244$

$\qquad\qquad\qquad = 2.16 \times 10^{-5}$ mol dm^{-3}

Hence $\qquad$ pH $\qquad = 4.66$

Comment

The pH is that of an *acidic buffer*. The absolute concentrations of the two components are not needed. It is the ratio of their concentrations that determines the buffer pH. Since the final volume of the solution is common to both components, the concentration ratio is the same as the mole ratio of the two components.

Example

Calculate the change in pH of the buffer solution in the above example after the addition of 10.0 cm^3 of a 0.10 mol dm^{-3} solution of HCl.

Method

The addition of acid shifts the equilibrium below to the *left*:

$$CH_3COOH(aq) \rightleftharpoons CH_3COO^-(aq) + H^+(aq)$$

with $[H^+] = K_a \dfrac{[CH_3COOH]}{[CH_3COO^-]}$

Decrease the number of moles of CH$_3$COO$^-$ by the number of moles of HCl added and increase the number of moles of CH$_3$COOH in proportion.

Answer

moles H^+ in 10.0 cm^3 of 0.10 mol dm^{-3} HCl $= \dfrac{0.10 \times 10.0}{1000} = 0.0010 \text{ mol}$

Initial moles of CH_3COO^-	$= 0.0225 \text{ mol}$	
Thus new moles of CH_3COO^-	$= (0.0225 - 0.0010)$	$= 0.0215 \text{ mol}$
Initial moles of CH_3COOH	$= 0.0280 \text{ mol}$	
Thus new moles of CH_3COOH	$= (0.0280 + 0.0010)$	$= 0.0290 \text{ mol}$

and $[H^+]$ $= K_a \times \dfrac{0.0290}{0.0215}$

$\qquad\qquad = 1.74 \times 10^{-5} \times 1.349 \quad = 2.347 \times 10^{-5} \text{ mol dm}^{-3}$

Hence pH = 4.63 when, previously, it was 4.66

The change in pH is –0.03 units.

Comment

Adding a substantial volume of a 0.10 mol dm^{-3} solution of HCl has only a slight effect on the pH, as is predicted for a buffer solution.

Applications of buffer solutions

In addition to laboratory use when standardising pH meter, mentioned above, buffer solutions have many additional laboratory uses in biological experiments on living systems. For example, the growth of bacteria for hospital tests is possible only in buffered systems; the waste products of growth, be they predominantly acidic or predominantly basic, can rapidly alter the pH of the growth medium to toxic levels, causing the bacteria to die.

A prime example of buffers in living systems is our own blood, which needs to be maintained at a pH close to 7.4 in healthy humans. The mechanism that maintains this pH is rather complex; it involves several aqueous buffer systems including the H_2CO_3/HCO_3^- and $H_2PO_4^-/HPO_4^{2-}$ pairs, and is enhanced by the buffering action of haemoglobin and other blood proteins.

The carbonic acid system acts through a linked system of chemical and physical equilibria:

$$H^+(aq) + HCO_3^-(aq) \rightleftharpoons H_2CO_3(aq) \rightleftharpoons H_2O(l) + CO_2(aq) \rightleftharpoons H_2O(l) + CO_2(g)$$

A straightforward application of Le Chatelier's principle down this sequence shows how an increase in acidity (a decrease in pH) in the blood is relieved by the formation of more aqueous H_2CO_3, relieved in turn by the production of more aqueous CO_2 and, finally, in the lungs, relieved by an increased breathing rate and the exhalation of more gaseous CO_2.

An inanimate example of a buffer system exists in the seas and oceans that surround us. These are maintained at a slightly alkaline level, at a pH of between 7.5 and 8.4, by a complicated system of buffers based on silicates and hydrogencarbonates. An enhanced knowledge of this system, which adjusts some of the level of CO_2 in our immediate environment, is clearly a vital tool in our understanding of global warming and climate change.

Examiners' Notes

In a clinical condition called *acidosis*, the blood acquires too high a concentration of CO_2 and the pH drops. In order to expel the excess CO_2, rapid breathing and consequent discomfort result.

3.4.4 Nomenclature and isomerism in organic chemistry

Naming organic compounds

Aliphatic compounds

The basic rules for naming aliphatic organic compounds were illustrated in *Collins Student Support Materials: Unit 1 – Foundation Chemistry*, section 3.1.5 and in *Collins Student Support Materials: Unit 2 – Chemistry in Action*, section 3.2.10. A few functional groups which appear in the present unit were not included previously; these groups are listed in Table 9.

Table 9
Some homologous series and functional groups

Homologous series	Suffix	Functional group	Example
esters	-oate	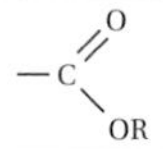 $-C\overset{O}{\underset{OR}{\diagdown}}$	ethyl ethanoate $CH_3COOCH_2CH_3$
acyl halides	-oyl halide	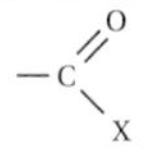 $-C\overset{O}{\underset{X}{\diagdown}}$	ethanoyl chloride CH_3COCl
acid anhydrides	-oic anhydride	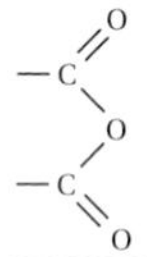 $-C\overset{O}{\underset{O}{\diagdown}}$ $-C\underset{O}{\diagup}$	ethanoic anhydride $(CH_3CO)_2O$

Aromatic compounds (see also section 3.4.6)

Benzene rings with only one substituent are usually straightforward to name, e.g. nitrobenzene or chlorobenzene. However, when a benzene ring has more than one substituent, the various groups are numbered in such a way that the lowest possible numbers are used. The substituents are *listed in alphabetical order*, together with their appropriate numbers. Some examples are shown in Fig 14.

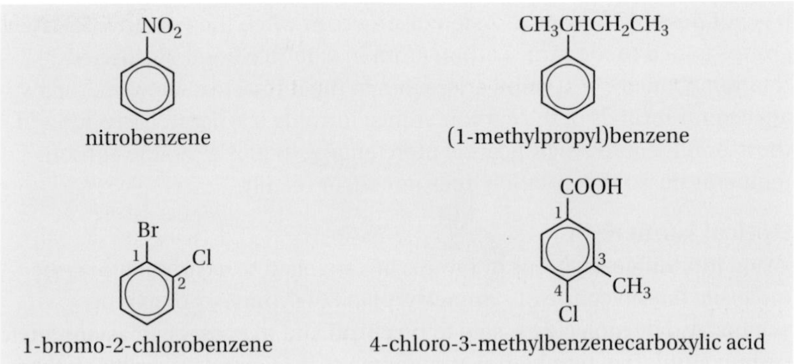

nitrobenzene

(1-methylpropyl)benzene $CH_3CHCH_2CH_3$

1-bromo-2-chlorobenzene

4-chloro-3-methylbenzenecarboxylic acid

Fig 14
Naming benzene derivatives

The phenyl group, C_6H_5, is sometimes regarded as a substituent, for example in $C_6H_5NH_2$, which is phenylamine, and $C_6H_5CH=CHCH_3$, which is 1-phenylpropene.

Isomerism

Structural isomerism was considered in *Collins Student Support Materials: Unit 1 – Foundation Chemistry*, section 3.1.5, and includes chain isomerism, position isomerism and functional group isomerism:

- **Chain isomerism** occurs when there are two or more ways of arranging the carbon skeleton of a molecule.

- **Position isomerism** occurs when the isomers have the same carbon skeleton, but the functional group is attached at different places on the chain.

- **Functional group isomerism** occurs when different functional groups are present in compounds which have the same molecular formula.

Stereoisomerism

The two types of stereoisomerism are **E–Z stereoisomerism** (see *Collins Student Support Materials: Unit 2 – Chemistry in Action*, section 3.2.9) and **optical isomerism**.

> **Definition**
>
> *Stereoisomers are compounds which have the same structural formula but the bonds are arranged differently in space.*

E–Z stereoisomerism

Because of restricted rotation at the $C=C$ bond, *Z* or *cis* and *E* or *trans* forms occur when there is suitable substitution:

Z or cis   E or trans

Essential Notes

E–Z stereoisomerism is also known as geometrical or *cis–trans* isomerism.

It is not possible to have *E–Z* stereoisomerism when there are two identical groups joined to the same carbon atom in a double bond. Restricted rotation about a $C=C$ bond arises due to the interaction between the two adjacent p-orbitals of the carbon atoms, forming a π-bond. Disruption of the π-bond requires significantly more energy than is available at room temperature, so that rotation does not occur readily.

Optical isomerism

When four different atoms or groups are attached to a carbon atom, the molecule has no centre of symmetry, plane of symmetry or axis of symmetry. The molecule is said to be **chiral** and to possess an **asymmetric** carbon atom. Two tetrahedral arrangements in space are possible so that one is the mirror image of the other; 2-hydroxypropanenitrile (see section 3.4.5) has two such stereoisomers (see Fig 15).

Stereoisomers of this kind are known as **enantiomers**. It is not possible to superimpose one enantiomer on the other. Enantiomers have the same physical properties except for their effect on the plane of plane-polarised light and, because of this difference, are said to be **optically active**.

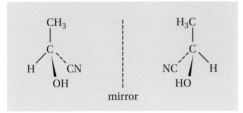

Fig 15
Stereoisomers of 2-hydroxypropanenitrile

When plane-polarised light, which is made up of waves vibrating in one plane only, passes through a solution of a chiral compound, the light emerges with its direction of polarisation changed. One enantiomer will rotate the plane of plane-polarised light in a clockwise direction; it is termed (+) or dextrorotatory. Its mirror-image form will rotate the plane of plane-polarised light by the same amount in a counterclockwise direction; it is termed (−) or laevorotatory. A mixture of equal amounts of both enantiomers is optically inactive, because the two effects cancel out. Such a mixture is called a **racemic mixture** or a **racemate**.

Many naturally-occurring molecules exist as single enantiomers, notably most amino acids (see section 3.4.8), such as 2-aminopropanoic acid, *alanine*, $CH_3CH(NH_2)COOH$. The chemical properties of enantiomers are identical except in reactions with other optically active substances. Because enzymes are stereospecific, they can distinguish between enantiomers and catalyse the reactions of only one of a pair of isomers.

2-hydroxypropanenitrile can be prepared (Fig 18, R = CH_3) by the reaction of hydrogen cyanide with ethanal (see section 3.4.5). Because the carbonyl group is planar, attack by the nucleophile CN^- is equally likely from either side of the plane, leading to the formation of a racemate (Fig 16). Hydrolysis of this product causes the CN group to be converted into COOH. The final product is the racemate of 2-hydroxypropanoic acid, commonly called *lactic acid*. This (±)-mixture is found in sour milk; the naturally occurring (+)-enantiomer is formed during the contraction of muscles.

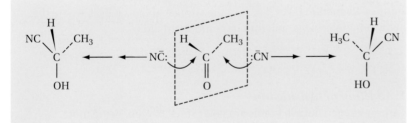

Fig 16
Racemate formation

Chiral drugs

Approximately half of the commercially-available drugs contain at least one chiral centre. Although drugs extracted from natural sources, e.g. quinine, are single enantiomers, most synthetic products are obtained as racemic mixtures. Because of the inherent difficulty in, and high cost of, separating the enantiomers, with few exceptions, synthetic drugs have

Essential Notes

The maximum number of stereoisomers for a compound with n chiral centres is 2^n; this number is reduced by symmetrical substitution.

been marketed as racemates. However, enantiomers can have unequal degrees of the same physiological activity or very different activities.

The drug thalidomide was used originally, as the racemate, to combat morning sickness in pregnant women. One enantiomer proved to be a potent teratogen (causes malformation of a fetus), leading to severe birth defects. The other enantiomer is not teratogenic but, unfortunately, in this particular case, the two stereoisomers can interconvert **in vivo** (within the human body).

The separation of enantiomers, known as **resolution**, can be achieved by reaction of the mixture with a chiral reagent, resulting in products with different physical properties, e.g. solubility. Enantiomers can also be separated directly by chromatography (see section 3.4.11).

3.4.5 Compounds containing the carbonyl group

Aldehydes and ketones

Oxidation of aldehydes

The oxidation of aldehydes to carboxylic acids was covered in *Collins Student Support Materials: Unit 2 – Chemistry in Action*, section 3.2.10:

$$RCHO + [O] \rightarrow RCOOH$$

That section described the use of Fehling's solution or Tollens' reagent to distinguish between aldehydes and ketones (see Table 14 on page 67).

Reduction of aldehydes and ketones

Reduction of the carbonyl group in aldehydes and ketones leads to the formation of primary alcohols and secondary alcohols, respectively.

1. Catalytic hydrogenation

In the presence of a metal catalyst, such as finely-dispersed nickel, hydrogen adds to the carbon–oxygen double bonds of aldehydes and ketones. Reduction takes place on the surface of the catalyst, where the hydrogen molecule is split into its component atoms:

$$RCHO + H_2 \rightarrow RCH_2OH$$

$$RCOR + H_2 \rightarrow RCH(OH)R$$

2. Addition of hydride ion

Carbon–oxygen double bonds are polar since oxygen, being more electronegative than carbon, has a greater share of the bonding electrons between the two atoms. Such groups are readily reduced by reagents which are sources of nucleophilic hydride ions ($:H^-$). Sodium tetrahydridoborate(III) ($NaBH_4$) can be used in aqueous ethanol, but anhydrous conditions are required for the more powerful reducing agent lithium tetrahydridoaluminate(III) ($LiAlH_4$).

Examiners' Notes

Note that reduction of an unsymmetrical ketone gives rise to a racemic product, e.g. $CH_3COCH_2CH_3$ is converted into $(\pm)$-$CH_3CH(OH)CH_2CH_3$.

Examiners' Notes

Note that $NaBH_4$ has no effect on alkenes, neither does $LiAlH_4$, so that selective reduction of the carbonyl group can be achieved in compounds containing both carbon–carbon and carbon–oxygen double bonds. A reducing agent such as hydrogen in the presence of a nickel or a platinum catalyst would reduce all unsaturated linkages.

The mechanism of hydride-ion reduction involves nucleophilic attack on the electron-deficient (δ+) carbon atom of the carbonyl group, to form an oxyanion which is subsequently protonated by water or a weak acid (Fig 17).

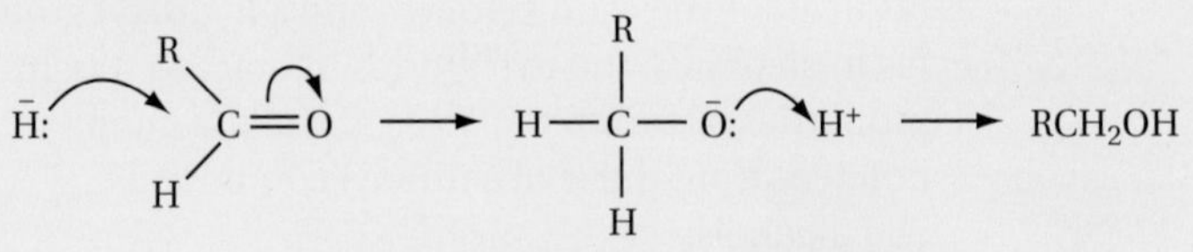

$RCHO + 2[H] \rightarrow RCH_2OH$

Fig 17
Hydride reduction of an aldehyde

3. Addition of hydrogen cyanide

Hydrogen cyanide adds nucleophilically to aldehydes and ketones to form hydroxynitriles, leading to an increase in the number of carbon atoms. Thus, ethanal is converted into 2-hydroxypropanenitrile:

$CH_3CHO + HCN \rightarrow CH_3CH(OH)CN$

Because hydrogen cyanide is a highly toxic gas, the best way of carrying out the reaction is to generate the HCN in the reaction mixture by adding a dilute acid to an excess of aqueous sodium cyanide.

In the mechanism, the nucleophilic cyanide ion attacks the carbonyl group to form an oxyanion which then accepts a proton (Fig 18).

Examiners' Notes

The use of NaCN or KCN avoids the danger of rapid death caused by inhalation of HCN. However, care is still required because these salts are very toxic if ingested. An aqueous solution of HCN itself is not a good source of cyanide ions ($pK_a = 9.40$).

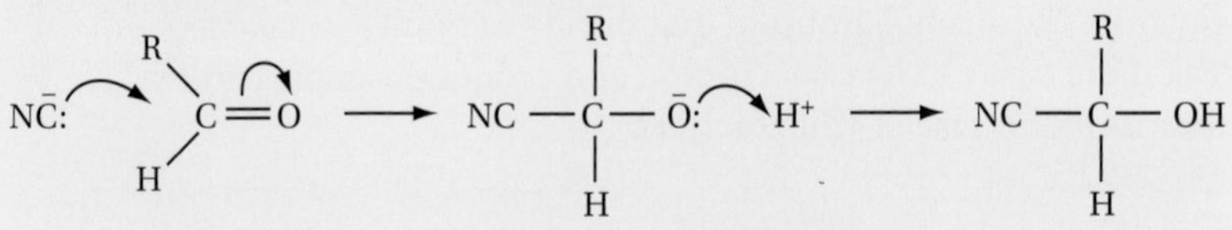

Fig 18
Nucleophilic addition of hydrogen cyanide

The reaction product is formed as an optically-inactive racemate, because attack by the cyanide nucleophile occurs with equal probability from either side of the planar carbonyl group (see Fig 16).

The addition of hydrogen cyanide to aldehydes and ketones is useful in synthesis. Catalytic hydrogenation reduces the CN group to the CH_2NH_2 group (see section 3.4.7). 2-Hydroxypropanenitrile is used to illustrate this type of reaction in Fig 19.

Essential Notes

The CN group is converted into the COOH group by acid-catalysed hydrolysis.

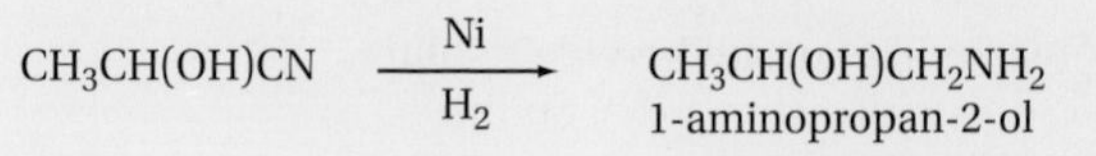

$CH_3CH(OH)CN \xrightarrow[H_2]{Ni} CH_3CH(OH)CH_2NH_2$
1-aminopropan-2-ol

Fig 19
Reduction of 2-hydroxypropanenitrile

Carboxylic acids and esters

Carboxylic acids

Carboxylic acids contain the functional group COOH

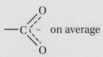

a carboxyl group

Although this group is made up of a **carb**onyl group (as in aldehydes and ketones) and a hyd**roxyl** group (as in alcohols), the two groups interact so that the properties of carboxylic acids are, in the main, different from those of carbonyl compounds and alcohols.

Table 10
Some simple carboxylic acids

Formula	Name
HCOOH	methanoic acid
CH_3COOH	ethanoic acid
CH_3CH_2COOH	propanoic acid
C_6H_5COOH	benzenecarboxylic acid

Acidity

Carboxylic acids of low relative molecular mass (low M_r) are very soluble in water because the COOH group forms hydrogen bonds with water. The solubility is much lower, however, if the group attached is a long-chain alkyl or an aryl substituent, when the influence of the carboxylic acid group on the overall physical properties is much reduced.

Although the solubility may be high, the acids are only slightly dissociated in water, i.e. they are weak acids. The carbonyl group attracts electrons away from the alcohol group so that the O—H bond is weakened and can break more easily to release a proton and produce a stable carboxylate anion, as in the case of ethanoic acid:

$$CH_3COOH(aq) \rightleftharpoons CH_3COO^-(aq) + H^+(aq)$$

As distinct from strong acids, the amount of this dissociation is quite small. The dissociation constant, K_a, for ethanoic acid is 1.76×10^{-5} mol dm^{-3}, which means that a solution containing 0.1 mol dm^{-3} of the acid is only about 0.3% ionised. By comparison, the O—H bond in an alcohol does not break easily, so that alcohols do not show typical acidic properties.

Carboxylic acids react as normal acids with metals, alkalis and carbonates to form salts, although the reactions are less vigorous than with strong acids.

When carboxylic acids react with sodium hydrogencarbonate, carbon dioxide is evolved. This reaction can be used as a test for carboxylic acids:

$$CH_3COOH + NaHCO_3 \rightarrow CH_3COONa + H_2O + CO_2$$

The salts formed are ionic and are therefore water-soluble.

Esters

Esters contain the functional group COOR, where R is usually an alkyl group. Esters can be formed by the reaction of carboxylic acids with alcohols in the presence of strong-acid catalysts, as in the case of methyl ethanoate (see also acylation):

$$CH_3COOH + CH_3OH \rightleftharpoons CH_3COOCH_3 + H_2O$$

When ethanoic acid and methanol are mixed in the presence of a small quantity of concentrated sulfuric acid, methyl ethanoate is formed as a volatile, sweet-smelling product.

Naming esters

The ionic carboxylate group in a salt of a carboxylic acid is named by substituting **-oate** for the -oic ending from the name of the carboxylic acid.

The acid part of an ester is similarly named:

CH_3COO^- is called ethanoate

$CH_3CH_2COO^-$ is called propanoate

The alcohol part of the ester name is written first, as in Table 11.

Name	Formula
methyl propanoate	$CH_3CH_2COOCH_3$
ethyl methanoate	$HCOOCH_2CH_3$
ethyl benzenecarboxylate	$C_6H_5COOCH_2CH_3$

Table 11
Examples of complete names

Uses of esters

1. Solvents

Esters have no free OH groups, so cannot form hydrogen bonds in the same way as carboxylic acids or alcohols; they are therefore much more volatile than carboxylic acids and are almost insoluble in water. Esters are polar, however, and are able to act as solvents for many polar organic compounds; their relatively low boiling points allow them to be separated easily from the less volatile solutes. Typical examples include:

- ethyl ethanoate (b.p. 77 °C), which is commonly used as the solvent in glue, such as polystyrene cement, and in printing inks;

- butyl ethanoate (b.p. 126 °C), which is widely used as a solvent in the pharmaceutical industry, for nitrocellulose, and in many lacquers.

2. Plasticisers

Although the forces between the chains in thermoplastic polymers are weak, the material is often not soft or flexible because the polymer chains cannot move easily over each other. Incorporating plasticisers into the polymer allows the chains some movement and, by adding different amounts of plasticiser, the flexibility of the material can be adjusted. For example, the esters of benzene-1,2-dicarboxylic (*phthalic*) acid or of hexanedioic (*adipic*) acid can constitute up to 50% of some plastics such as PVC (see section 3.4.9). Over time these additives escape from the plastic, which then becomes brittle and stiff with age.

3. Food flavourings

Many esters have sweet, often fruity, smells and are used as artificial fruit flavourings. Some of the common esters used in flavourings are shown

Essential Notes

Ester formation involves the elimination of water. Isotopic labelling with ^{18}O shows that the OH group is lost from the acid and not the alcohol.

When making esters from carboxylic acids, carefully pour the equilibrium mixture formed into an excess of warm water. The remaining carboxylic acid and alcohol dissolve, but the ester is immiscible and floats on the water. The aroma of the ester is no longer contaminated with that of the acid and is easily detected.

Examiners' Notes

It should be recognised that esters and carboxylic acids are good examples of functional group isomerism (see *Collins Student Support Materials: Unit 1 – Foundation Chemistry*, section 3.1.5).

Examiners' Notes

The addition of suitable plasticisers to PVC enables the resulting material to be used as a synthetic leather.

in Table 12. Natural fruit flavours are extremely complex mixtures of many esters and carboxylic acids. Artificial flavours using only some of these compounds will inevitably only approximate to the real thing.

Table 12
Esters used as food flavourings

Name	Flavour
pentyl ethanoate	pear
2,2-dimethylpropyl ethanoate	banana
octyl ethanoate	orange
ethyl butanoate	rum
pentyl pentanoate	apple

Essential Notes

Solid esters occur in animal fats; vegetable oils are liquid esters.

Examiners' Notes

Whereas the sodium salts of long-chain carboxylic acids are soluble in water, the calcium salts are not. Hence, in hard-water areas, the calcium ions form an unsightly scum with soap. Detergents, however, are sodium salts of sulfonic acids, which form soluble calcium salts, so that no scum is formed when a detergent is used:

R = unbranched alkyl chain of about 12 carbon atoms

Hydrolysis of esters

When an ester is heated with alkali, it is hydrolysed to an alcohol and a carboxylate salt, e.g. ethyl ethanoate is hydrolysed by aqueous sodium hydroxide to produce ethanol and sodium ethanoate:

$$CH_3COOC_2H_5 + NaOH \rightarrow C_2H_5OH + CH_3COONa$$

This hydrolysis reaction is used widely with naturally-occurring esters, such as oils and fats, to produce useful products including soaps and glycerol.

Most oils and fats are esters of propane-1,2,3-triol (*glycerol*) with three long-chain carboxylic acids. These acids are often called fatty acids and the esters are called triglycerides. The most common fatty acids are:

- octadeca-9-enoic (*oleic*) acid, an unsaturated acid which occurs in most fats and in olive oil;

- octadecanoic (*stearic*) acid, a saturated acid which occurs in animal fats;

- octadeca-9,12-dienoic (*linoleic*) acid, an unsaturated compound which is the principal acid in many vegetable oils, such as soya bean and corn oil.

When fats are boiled with sodium hydroxide, glycerol is formed together with a mixture of the sodium salts of the three component acids. These salts are soaps:

$$
\begin{array}{l}
CH_2OOC(CH_2)_{16}CH_3 \\
| \\
CHOOC(CH_2)_{16}CH_3 \quad + \quad 3NaOH \quad \rightarrow \\
| \\
CH_2OOC(CH_2)_{16}CH_3
\end{array}
\qquad
\begin{array}{l}
CH_2OH \\
| \\
CHOH \quad + \quad 3CH_3(CH_2)_{16}COONa \\
| \\
CH_2OH
\end{array}
$$

propane-1,2,3-triol   sodium octadecanoate (a soap)

In practice, most soaps are mixtures of salts of long-chain carboxylic acids, since the fats and oils from which they are formed contain mixtures of these acids.

Glycerol, with its three O—H bonds, forms hydrogen bonds very easily and so has many applications that depend on its ability to attract water; it is used in the cosmetics industry, in food and in glues (to prevent materials drying too quickly). Glycerol is also an important component of wine, adding both to the sweetness and to the viscosity.

Biodiesel

Biodiesel is a renewable, non-petroleum-based fuel obtained mainly from vegetable oils by acid- or base-catalysed **transesterification**. Soya bean and rapeseed oils are most commonly used as biodiesel feedstocks. Waste vegetable oils and discarded animal fats are also used, but to a much lesser extent.

The most common form of biodiesel is a mixture of methyl esters of long-chain fatty acids, such as that formed with propane-1,2,3-triol in the following reaction:

$$
\begin{array}{c}
CH_2OOCR^1 \\
| \\
CHOOCR^2 \\
| \\
CH_2OOCR^3
\end{array}
\ + \ 3CH_3OH \ \rightleftharpoons \
\begin{array}{c}
CH_2OH \\
| \\
CHOH \\
| \\
CH_2OH
\end{array}
\ + \
\begin{array}{c}
R^1COOCH_3 \\
R^2COOCH_3 \\
R^3COOCH_3
\end{array}
$$

Biodiesel can be used alone (B100) or blended with petrodiesel. A fuel that can be used in unmodified diesel engines, containing 20% biodiesel and 80% petrodiesel, is labelled B20.

Biodiesel is non-toxic and biodegradable. It is cleaner burning than petrodiesel. New uses for the glycerol by-product are being explored. Biodiesel production capacity is growing rapidly and reached 10 million tonnes in Europe during 2007.

There is a danger that farmers might stop planting staple food crops in favour of those that produce biofuels, leading to a shortage of food.

Acylation

Acyl chlorides and acid anhydrides are useful synthetic intermediates for the preparation of other compounds. Both of these carboxylic acid derivatives possess good leaving groups (Cl^- and ^-OCOR, respectively) which also activate the adjacent carbonyl group by electron-withdrawal. Consequently, the carbon atom of the carbonyl group is susceptible to nucleophilic attack by water (hydrolysis) to give the acid, by alcohols to form esters, by ammonia to obtain amides and by amines to produce *N*-substituted amides (Fig 20). Although less vigorous, reactions of acid anhydrides with nucleophiles are analogous to those of acyl chlorides. Except in the case of hydrolysis, use of acid anhydrides results in the formation of co-products which are not easily removed.

$$
\begin{aligned}
RCOCl + H_2O &\rightarrow RCOOH + HCl \\
(RCO)_2O + H_2O &\rightarrow 2RCOOH \\
&\qquad\quad \text{acid} \\
RCOCl + CH_3OH &\rightarrow RCOOCH_3 + HCl \\
(RCO)_2O + CH_3OH &\rightarrow RCOOCH_3 + RCOOH \\
&\qquad\quad \text{ester} \\
RCOCl + 2NH_3 &\rightarrow RCONH_2 + NH_4Cl \\
(RCO)_2O + 2NH_3 &\rightarrow RCONH_2 + RCOO^-NH_4^+ \\
&\qquad\quad \text{amide} \\
RCOCl + 2CH_3NH_2 &\rightarrow RCONHCH_3 + CH_3NH_3^+Cl^- \\
(RCO)_2O + 2CH_3NH_2 &\rightarrow RCONHCH_3 + RCOO^-CH_3NH_3^+ \\
&\qquad\quad \textit{N}\text{-substituted amide}
\end{aligned}
$$

Examiners' Notes

Transesterification is a reversible reaction in which an ester reacts with an alcohol, usually in excess, to form a new ester and a new alcohol.

Note that polyesters can be manufactured by transesterification (see section 3.4.9).

Essential Notes

Note that a **good leaving group** is a stable species which is liberated during a chemical reaction.

Examiners' Notes

Acyl chlorides are reactive derivatives of carboxylic acids and can be obtained from the acids by reaction with sulfur dichloride oxide (*thionyl chloride*), as shown for ethanoyl chloride:

$CH_3COOH + SOCl_2 \rightarrow$
$\qquad CH_3COCl + SO_2 + HCl$

Examiners' Notes

Acid anhydrides are produced when acyl chlorides react with carboxylic acid salts:

$RCOONa + RCOCl \rightarrow$
$\qquad (RCO)_2O + NaCl$

Fig 20
Acylation reactions

The mechanism of each of the above reactions involves nucleophilic addition to the carbonyl group, followed by elimination, and can be illustrated by specific examples (Fig 21 to Fig 24).

Essential Notes

Note that the mechanisms of reactions involving ethanoic anhydride do not form part of the AQA A-level specification.

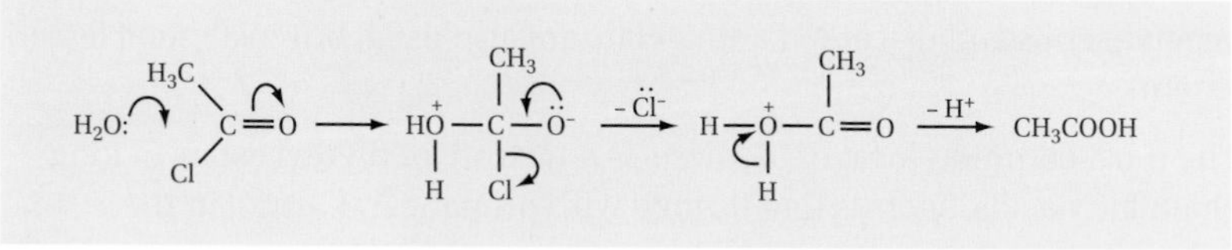

Fig 21
Hydrolysis of ethanoyl chloride

Examiners' Notes

Unlike the reaction between ammonia and haloalkanes, further acylation is difficult because the lone pair on the amide nitrogen atom is withdrawn by the carbonyl group, so that the amide is less likely to act as a nucleophile:

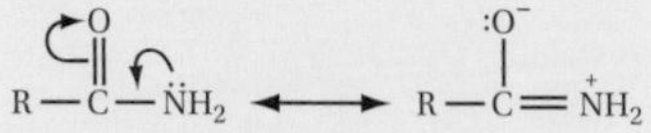

Consequently, amides are much less basic than amines (see section 3.4.7).

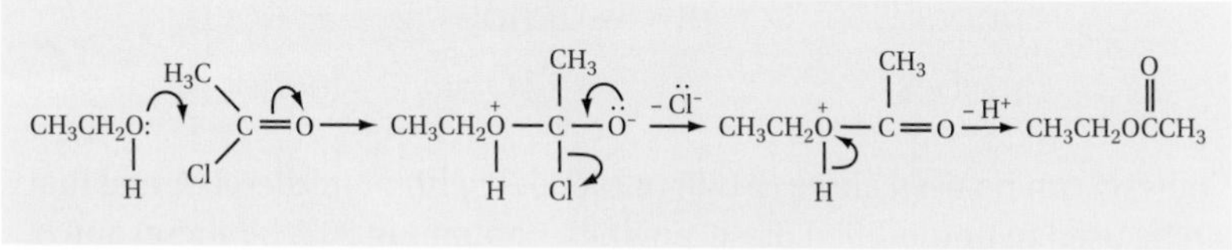

Fig 22
Formation of ethyl ethanoate

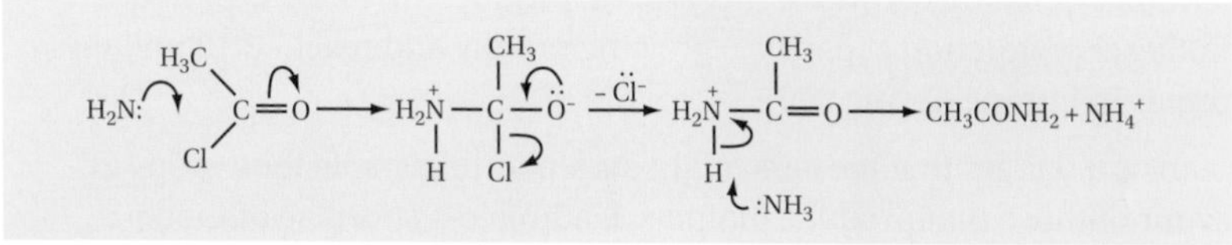

Fig 23
Formation of ethanamide (*acetamide*)

Fig 24
Formation of
N-phenylethanamide
(*acetanilide*)

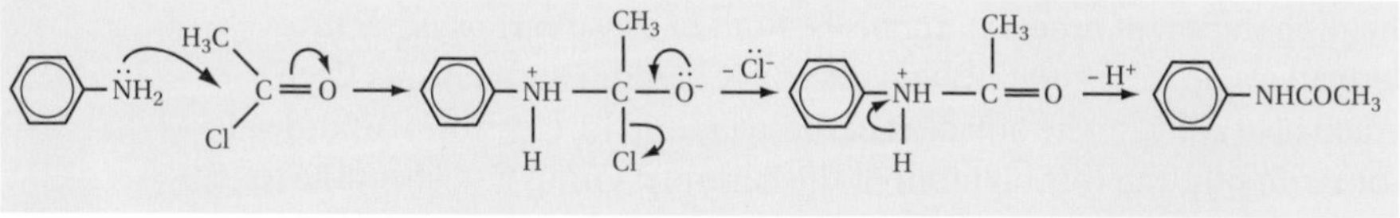

Examiners' Notes

Note that the mechanism of this type of acylation differs from that discussed in section 3.4.6.

The reaction between acyl chlorides and alcohols (Fig 22) is a highly effective way of producing esters. A base is usually added to neutralise the liberated hydrochloric acid. This method of preparation then avoids the equilibrium problem encountered in acid-catalysed esterification (see previous section). In the case of amide formation (Fig 23), an excess of ammonia is used to neutralise the liberated hydrochloric acid. Acylation is used to form derivatives of both aliphatic and aromatic amines (Fig 24).

Uses of acylation reactions

Unlike ethanoyl chloride, ethanoic anhydride is manufactured on a large scale for use as an acylating agent. The acid anhydride is relatively cheap compared with the acid chloride and is also less corrosive, less vulnerable to hydrolysis and less dangerous to use. A major use for ethanoic anhydride is in the manufacture of 2-ethanoyloxybenzenecarboxylic acid (*aspirin*).

Essential Notes

Note that hydroxybenzene derivatives are called phenols.

Aspirin is probably the most widely used drug of all time, being mainly employed as an analgesic (pain killer). The phenolic hydroxy group in the starting material, 2-hydroxybenzenecarboxylic (*salicylic*) acid, is acylated by heating the salicylic acid with ethanoic anhydride (Fig 25).

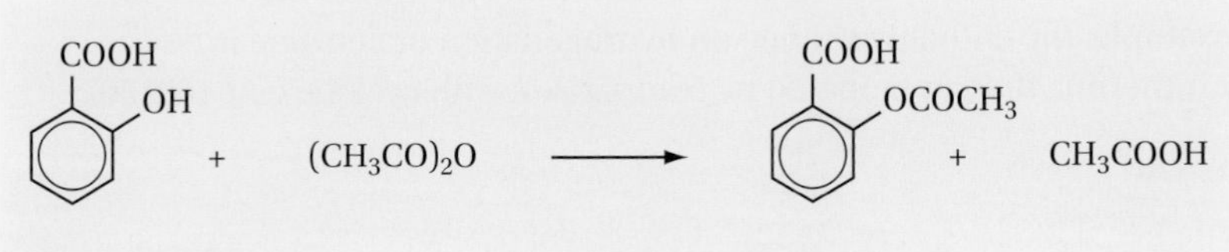

Fig 25
Synthesis of 2-ethanoyloxybenzenecarboxylic acid (*aspirin*)

3.4.6 Aromatic chemistry

Benzene, C_6H_6, is the parent of a group of cyclic unsaturated compounds known as **arenes**. Traditionally, such compounds are referred to as being **aromatic** because many naturally-occurring fragrant substances were found to contain substituted benzene rings.

Bonding

The first satisfactory formula for benzene was put forward in 1865 by Kekulé, who suggested that the molecule had a cyclic arrangement of carbon atoms joined together by alternate single and double bonds. Two equivalent hexagonal structures can be drawn, so that isomeric 1,2-disubstituted benzenes might be expected:

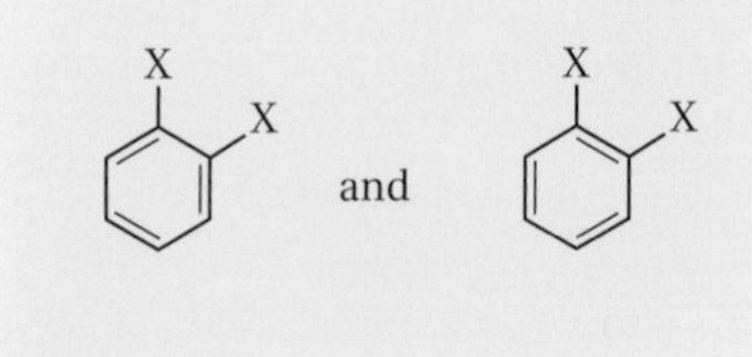

In fact, only a single 1,2-disubstituted benzene exists. This observation was explained by Kekulé, who proposed that a rapid equilibrium existed between the two equivalent structures, thereby averaging out the single and double bonds.

Single-crystal X-ray diffraction analysis has shown that the benzene molecule is a planar, regular hexagon in which the carbon–carbon bond lengths are all the same (139 pm), being intermediate between normal single bonds (154 pm) and double (134 pm) bonds. The identical bonding between carbon atoms in benzene is implied by the use of a circle inside the hexagon to indicate six **delocalised** electrons:

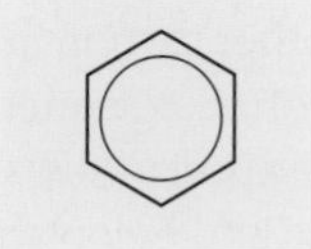

Examiners' Notes

Benzene is often considered to be a **resonance hybrid** of the two Kekulé structures, neither of which actually exists:

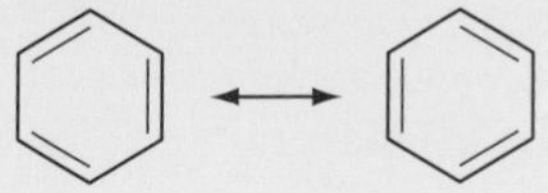

Examiners' Notes

Care has to be taken when the circle notation is applied to polycyclic molecules, such as naphthalene:

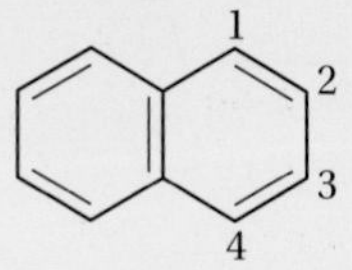

This fused bicyclic arene has a total of *ten* p-electrons, not twelve, and should **not** be shown with circles in each of the two rings. It is not possible for both rings to be benzenoid at the same time and the 1,2-bond is shorter than the 2,3-bond.

Delocalisation stability

Thermochemical evidence shows that benzene is much more stable than the hypothetical cyclohexa-1,3,5-triene molecule would be. Thus, for example, the enthalpy change on hydrogenation of benzene is less exothermic than anticipated by comparison with cyclohexene (Fig 26).

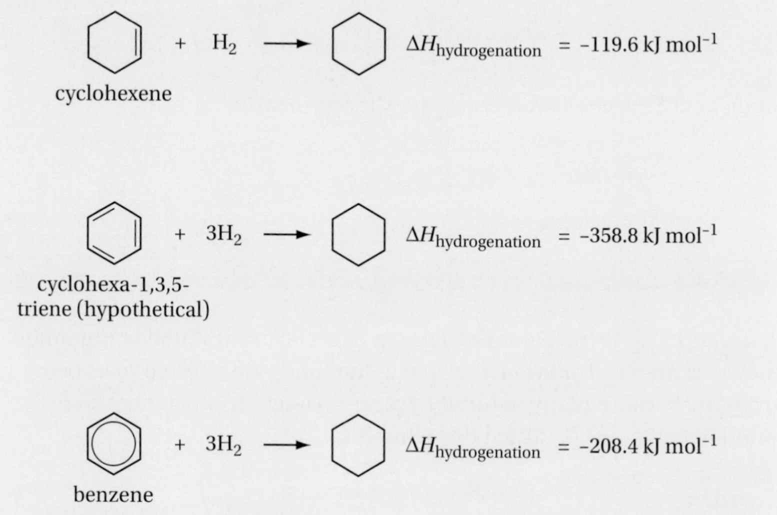

cyclohexene $+ H_2 \longrightarrow$ $\Delta H_{hydrogenation} = -119.6\ \text{kJ mol}^{-1}$

cyclohexa-1,3,5-triene (hypothetical) $+ 3H_2 \longrightarrow$ $\Delta H_{hydrogenation} = -358.8\ \text{kJ mol}^{-1}$

benzene $+ 3H_2 \longrightarrow$ $\Delta H_{hydrogenation} = -208.4\ \text{kJ mol}^{-1}$

Fig 26
Enthalpies of hydrogenation

Assuming the enthalpy of hydrogenation of cyclohexa-1,3,5-triene to be three times that of cyclohexene gives the value of $-358.8\ \text{kJ mol}^{-1}$. On this basis, benzene is $150.4\ \text{kJ mol}^{-1}$ more stable than cyclohexa-1,3,5-triene. These differences are illustrated diagrammatically in Fig 27.

Fig 27
Hydrogenation of cyclohexene and benzene

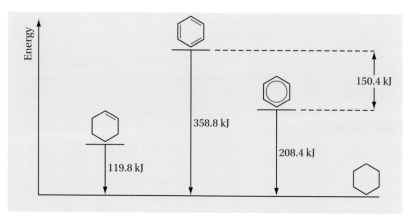

Examiners' Notes

You can see from this diagram that benzene is $150.4\ \text{kJ mol}^{-1}$ **more** stable than the hypothetical molecule cyclohexa-1,3,5-triene.

All the bonds in benzene are identical because of the electronic structure of the molecule: each of the six carbon atoms in the hexagonal arrangement is bonded to two other carbon atoms and to one hydrogen atom, and all bond angles are 120°. Consequently, each carbon atom has one unused p-electron. Delocalisation of the six p-electrons gives rise to regions of electron density above and below the plane of the ring.

This cyclic electron delocalisation has a marked stabilising effect so that benzene undergoes overall addition reactions with difficulty. The increase in stability associated with electron delocalisation is called the **delocalisation energy** or the **resonance energy**.

Electrophilic substitution

The benzene molecule is susceptible to attack by positively-charged species: **electrophiles**. Although the first step in the substitution reaction involves addition of the electrophile (E^+), the resulting cationic intermediate then loses a proton to regenerate the delocalised (stable) system, resulting in the formation of a substitution product (Fig 28).

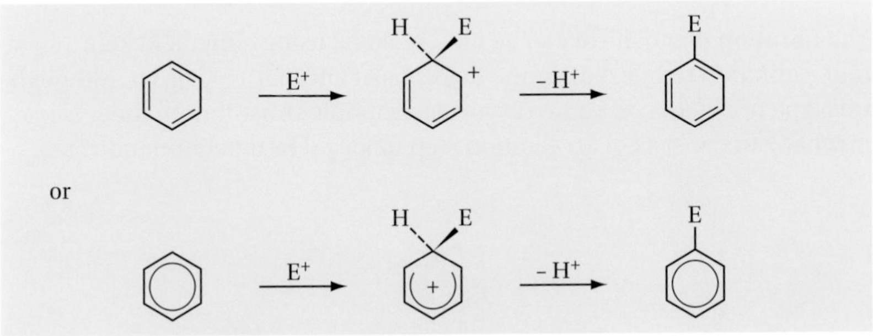

Essential Notes

Addition to benzene can occur, e.g. hydrogenation to form cyclohexane, but only with difficulty because the delocalisation energy is lost.

Examiners' Notes

Note that either of these representations of electrophilic substitution is acceptable in examination answers. The second representation is discussed in more detail on page 55.

Fig 28
Electrophilic substitution of benzene

Electrophilic substitution reactions such as nitration and acylation are important steps in synthesis. In these reactions it is first necessary to form an electrophile, which then attacks the benzene ring.

Nitration

In nitration reactions, the electron-deficient **nitryl cation**, NO_2^+ (*nitronium ion*), is the electrophile. This species is generated via the protonation of nitric acid by a stronger acid, usually concentrated sulfuric acid. The overall equation:

$$HNO_3 + 2H_2SO_4 \rightleftharpoons {}^+NO_2 + H_3O^+ + 2HSO_4^-$$

summarises the three equilibria shown in Fig 29.

Essential Notes

Nitronium salts such as the tetrafluoroborate ($NO_2^+BF_4^-$) can also be used as nitrating agents.

$$HNO_3 + H_2SO_4 \rightleftharpoons [H_2NO_3]^+ + HSO_4^-$$

$$[H_2NO_3]^+ \rightleftharpoons H_2O + {}^+NO_2$$

$$H_2O + H_2SO_4 \rightleftharpoons H_3O^+ + HSO_4^-$$

Examiners' Notes

Note that, in the first equilibrium, HNO_3 is acting as a base.

Fig 29
Generation of the nitronium ion

Breakdown of the protonated nitric acid molecule gives water and the nitronium ion (Fig 30).

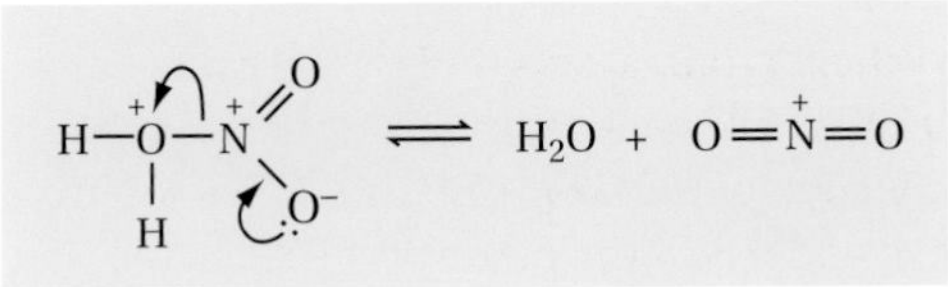

Fig 30
Formation of $^+NO_2$

Nitrobenzene is obtained by warming benzene with a mixture of concentrated nitric acid and concentrated sulfuric acid (Fig 31). The sulfuric acid aids the formation of the inorganic electrophile by the removal of water (Fig 29).

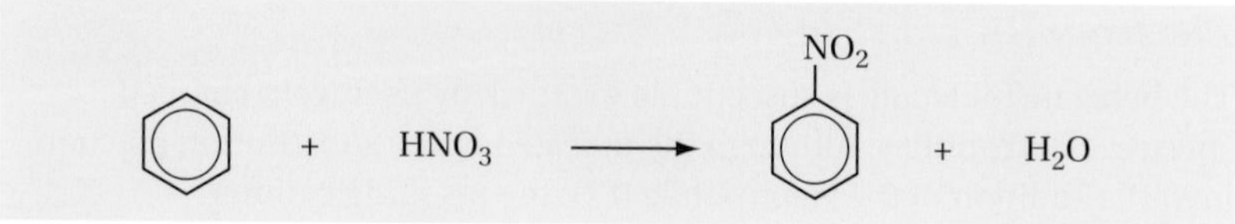

Fig 31
Nitration of benzene

The nitration mechanism can be illustrated by using either a Kekulé representation of benzene or the delocalised alternative (Fig 32). Although this type of reaction is known as an electrophilic substitution, the mechanism consists of an addition step followed by an elimination step.

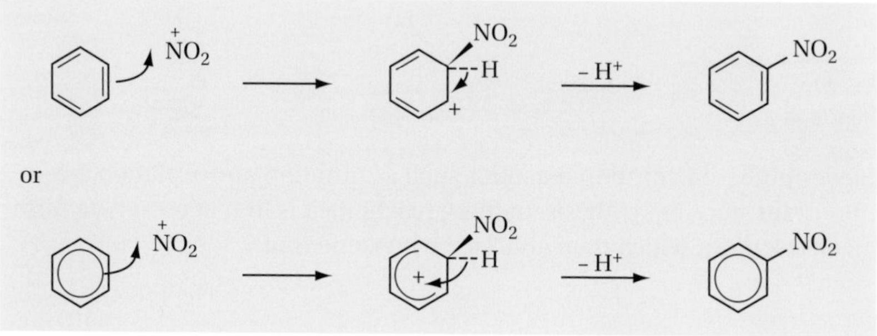

Fig 32
Mechanism of nitration

Use of nitration reactions

Aromatic nitro compounds are important as precursors of aromatic amines. The reduction of nitro compounds can be achieved easily by catalytic hydrogenation (e.g. Ni/H_2) or by using metal/acid combinations (e.g. Sn/HCl) (see section 3.4.7). The resulting primary aromatic amines are very useful intermediates in organic synthesis.

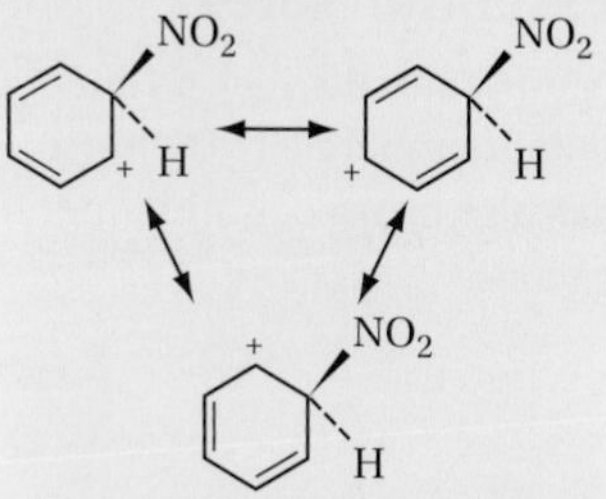

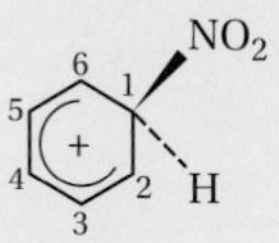

Friedel–Crafts acylation reactions

Friedel–Crafts reactions are important examples of electrophilic substitution because they lead to carbon–carbon bond formation. In 1877, Friedel and Crafts discovered that a haloalkane will react with benzene in the presence of a catalyst such as aluminium chloride (Fig 33).

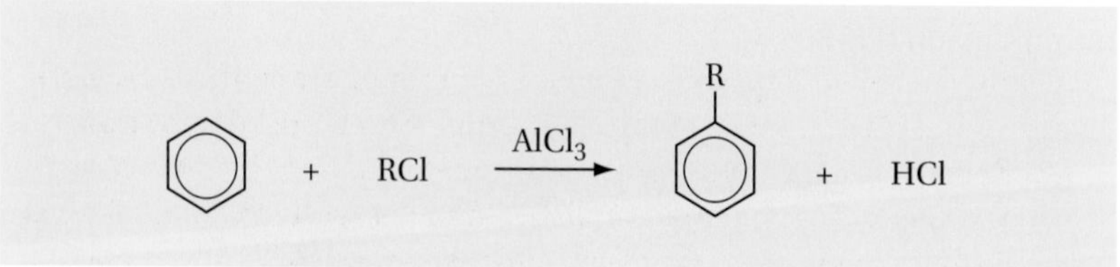

Fig 33
Friedel–Crafts alkylation of benzene

Alkylation reactions do not constitute important steps in synthesis, unlike corresponding acylation reactions which introduce a reactive functional group into the aromatic ring. Thus, ketones are produced when benzene reacts with acyl chlorides, in the presence of $AlCl_3$. The **acylium cation** initially formed is sufficiently electrophilic to attack benzene by the usual aromatic substitution mechanism (Fig 34).

$$RCOCl + AlCl_3 \rightarrow RCO^+ + AlCl_4^-$$

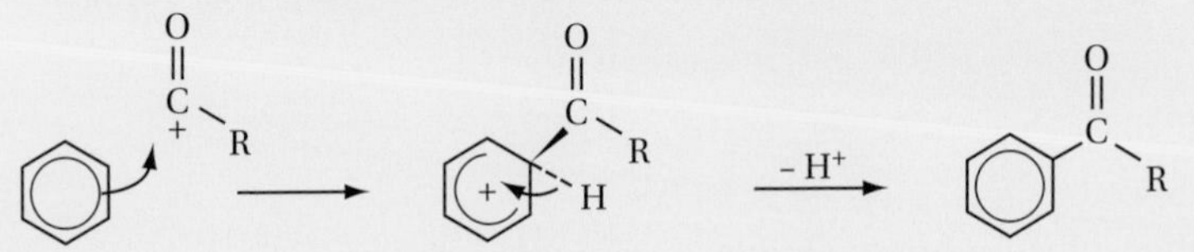

Fig 34
Mechanism of acylation

The formation of phenylethanone provides a specific example:

$$C_6H_6 + CH_3COCl \rightarrow C_6H_5COCH_3 + HCl$$

Carboxylic acid anhydrides are sometimes used instead of acyl chlorides. The acylium ion is formed in the presence of aluminium chloride:

$$(RCO)_2O + AlCl_3 \rightarrow RCO^+ + RCOOAlCl_3^-$$

Examiners' Notes

Only Friedel–Crafts acylation reactions are included in the AQA A-level specification.

Essential Notes

Acylium ions are often formed as major fragments in a mass spectrometer (see section 3.4.11).

Examiners' Notes

Note that the mechanism of this type of acylation differs from that discussed in section 3.4.5.

3.4.7 Amines

Replacement of the hydrogen atoms in ammonia by alkyl or aryl groups gives rise to three types of amine:

RNH_2	R_2NH	R_3N
primary	secondary	tertiary

The most important are primary amines, e.g. CH_3NH_2 (aliphatic) or $C_6H_5NH_2$ (aromatic). Boiling points are lower than those of the analogous alcohols, e.g. CH_3NH_2 is gaseous, because of weaker hydrogen bonding.

Base properties (Brønsted–Lowry)

Ammonia and amines act as Brønsted–Lowry bases (proton acceptors) by virtue of the lone pair of electrons on the nitrogen atom. The basicity is related to the availability of the lone pair (electron density) for protonation:

$$R\ddot{N}H_2 \; H^+ \rightleftharpoons RNH_3^+$$

The introduction of one alkyl group into ammonia strengthens the base due to the inductive effect of the alkyl group, which pushes electrons towards the nitrogen atom. For example, methylamine (pK_a 10.6) is a stronger base than ammonia (pK_a 9.2). On the other hand, arylamines are significantly weaker bases than alkylamines because of involvement of the

Examiners' Notes

K_a values for bases refer to dissociation of the conjugate acid:

$$RNH_3^+ \rightleftharpoons RNH_2 + H^+$$

so that the larger the K_a value (equilibrium to the right), the smaller the pK_a value and the weaker the base.

lone pair in the aromatic delocalisation, leading to a decrease in electron density on the nitrogen atom. Thus, for example, phenylamine (pK_a 4.6) is a weaker base than ammonia.

Nucleophilic properties

Essential Notes

Ligands are also lone-pair donors (see *Collins Student Support Materials: Unit 5 – Energetics, Redox and Inorganic Chemistry*, section 3.5.5).

Ammonia and amines also act as nucleophiles (electron-pair donors) and take part in nucleophilic substitution reactions. The reaction between a haloalkane and ammonia is an alkylation producing an alkylammonium salt. Proton exchange with another ammonia molecule produces the primary amine:

$$NH_3 + RBr \rightarrow [RNH_3]^+Br^-$$

$$[RNH_3]^+Br^- + NH_3 \rightleftharpoons \underset{\text{primary}}{RNH_2} + [NH_4]^+Br^-$$

Further substitution is possible, because the primary amine can compete effectively with ammonia for the haloalkane to generate a dialkylammonium salt. Further proton exchange with either ammonia or with RNH_2 liberates the secondary amine:

$$RNH_2 + RBr \rightarrow [R_2NH_2]^+Br^-$$

$$[R_2NH_2]^+Br^- + NH_3 \rightleftharpoons \underset{\text{secondary}}{R_2NH} + [NH_4]^+Br^-$$

Essential Notes

Note that a quaternary ammonium ion with four different groups is chiral.

A third alkylation can then take place to give a trialkylammonium salt which, in turn, will donate a proton to ammonia or to another amine:

$$R_2NH + RBr \rightarrow [R_3NH]^+Br^-$$

$$[R_3NH]^+Br^- + NH_3 \rightleftharpoons \underset{\text{tertiary}}{R_3N} + [NH_4]^+Br^-$$

Examiners' Notes

Surfactants (derived from **surface acting agents**) are wetting agents that lower the surface tension of a liquid and the interfacial tension between two liquids. Such chemicals contain both hydrophobic and hydrophilic groups and are soluble both in organic solvents and in water.

The resulting tertiary amine then reacts with the haloalkane in a fourth alkylation step to form a quaternary ammonium salt:

$$R_3N + RBr \rightarrow \underset{\text{quaternary ammonium salt}}{[R_4N]^+Br^-}$$

A mixture of products is usually obtained. Clearly, this outcome limits the usefulness of direct alkylation in synthesis, although separation of the various components is possible. A high yield of the quaternary ammonium salt is obtained by using a large excess of haloalkane. On the other hand, a large excess of ammonia reduces the possibility of further substitution and gives a better yield of primary amine.

Each of the four steps in the above sequence is mechanistically similar and involves nucleophilic attack from the side opposite the leaving group, as illustrated in Fig 35.

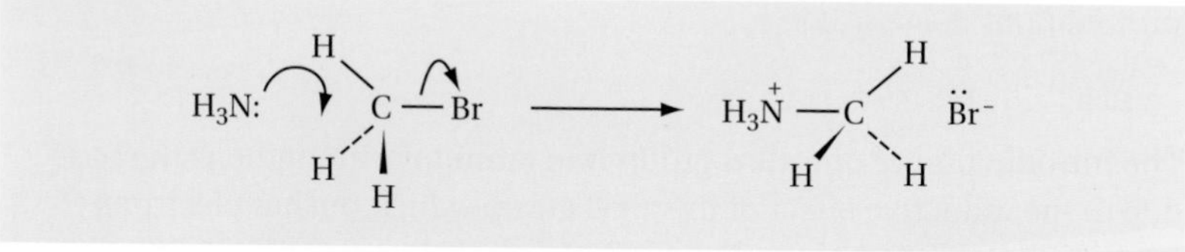

Fig 35
Formation of methylammonium bromide

Quaternary ammonium salts possessing two long-chain alkyl groups, such as $[CH_3(CH_2)_{17}]_2N(CH_3)_2{}^+Cl^-$, are used as cationic surfactants in fabric softening.

Preparation

A useful general method for preparing primary aliphatic amines of the type RCH_2NH_2 involves a two-step synthesis with a haloalkane as the starting material. Nucleophilic substitution with cyanide ion in aqueous ethanol gives the corresponding nitrile, which can then be reduced to a primary amine:

$$RBr + CN^- \rightarrow RC\equiv N + Br^-$$

$$RC\equiv N + 2H_2 \rightarrow RCH_2NH_2$$

Catalytic hydrogenation (e.g. Ni/H_2) is often used for the reduction as is lithium tetrahydridoaluminate(III) ($LiAlH_4$) in dry ethoxyethane.

Primary aromatic amines are usually prepared by the reduction of nitro compounds:

$$ArNO_2 + 3H_2 \xrightarrow{Ni} ArNH_2 + 2H_2O$$

Again, catalytic hydrogenation is a suitable method, giving almost quantitative (100%) yields. Methods of chemical reduction include heating with Sn/HCl (laboratory) or with Fe/HCl (industry). In these cases, the resulting amine is present in solution as $[ArNH_3]^+$. Consequently, a base (e.g. NaOH) is added to liberate the free amine. The organic product is then removed from the reaction mixture by distillation.

Examiners' Notes

Note that $NaBH_4$ is not sufficiently powerful to reduce a CN group.

Essential Notes

Just as R is often used to represent any alkyl group, Ar is similarly used to indicate an aryl group.

Examiners' Notes

Many functional groups can be introduced into benzene via the conversion of $ArNH_2$ into ArN_2^+ in a process called **diazotisation**.

3.4.8 Amino acids

The term **amino acid** is commonly used for compounds which have a primary amino group attached to the carbon atom adjacent to a carboxylic acid group:

$$RCH(NH_2)COOH$$

The position next to the carboxyl functional group is sometimes called the α-position, and a more complete name for such compounds is α-aminocarboxylic acids.

There are 20 naturally-occurring amino acids, which differ only in the nature of the group R. Apart from the simplest example, aminoethanoic acid, *glycine*, in which R = H, these compounds show optical activity. However, only one of the enantiomers occurs naturally. The amino acids are usually called by their common or trivial names. Some examples are given in Fig 36.

$CH_3CH(NH_2)COOH$	$HOCH_2CH(NH_2)COOH$
alanine	serine
$-CH_2CH(NH_2)COOH$	$H_2NCOCH_2CH(NH_2)COOH$
phenylalanine	asparagine

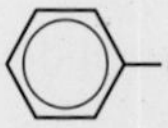

Fig 36
Examples of α-aminocarboxylic acids

Acid and base properties

Every amino acid has a carboxyl group and an amino group. So, depending on the pH of the solution in which the compound is dissolved, each group can exist in an ionic or a non-ionic form. The carboxyl groups of amino acids have pK_a values of about 2, whereas the protonated amino groups have pK_a values close to 9. Thus, in a very acidic solution, the amino group is protonated and the carboxyl group is undissociated. Conversely, in a strongly basic solution, the carboxyl group is deprotonated and the amino group is unchanged. In neutral solution, the carboxyl group is deprotonated and the amino group is protonated (Fig 37).

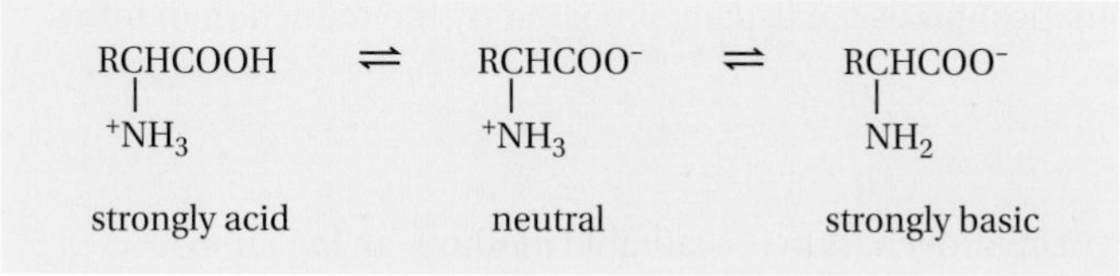

Fig 37
Amino acids in solution

RCHCOOH	$\rightleftharpoons$	RCHCOO$^-$	$\rightleftharpoons$	RCHCOO$^-$
$^+NH_3$		$^+NH_3$		NH_2
strongly acid		neutral		strongly basic

It is important to realise that an amino acid can never exist as an uncharged compound, regardless of the pH of the solution. At a pH of 7.3, found in living systems, an amino acid exists as a dipolar ion, $RCH(NH_3^+)COO^-$, which has an overall neutral charge. Such a species is called a **zwitterion**.

The zwitterionic nature of amino acids is reflected in the relatively high melting points of the crystals. Thus, glycine itself melts at about 290 °C, whereas the corresponding hydroxyethanoic acid, $HOCH_2COOH$, which is extensively hydrogen bonded but not ionic, melts at 80 °C.

Proteins

Proteins are naturally-occurring polymers of amino acids joined together by amide bonds (Fig 38) (see section 3.4.9).

Fig 38
Linking together amino acids

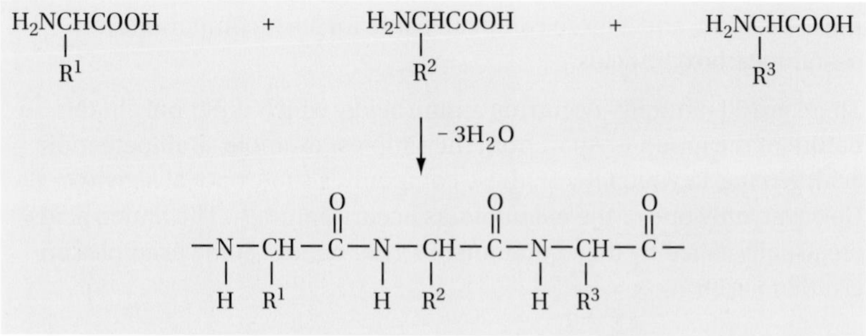

The amide bond (CO–NH) in proteins is often called a **peptide link**. Peptide is the term used to describe sequences of relatively few amino acids. For example, a tripeptide is made from three amino-acid units. Proteins can be thought of as complex, naturally-occurring polypeptides, which are made up of about 40 to around 4000 amino-acid units.

Hydrolysis of the peptide link

Peptides and proteins can be hydrolysed to the constituent amino acids in the presence of a strong acid or by the action of a specific enzyme catalyst. Fission of the peptide bonds in a protein liberates the component amino acids (Fig 39).

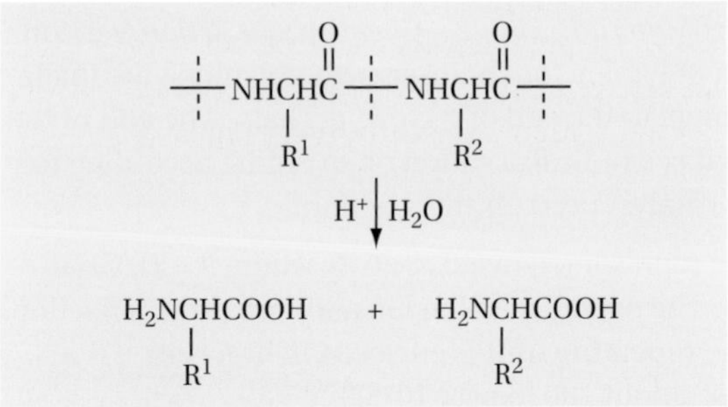

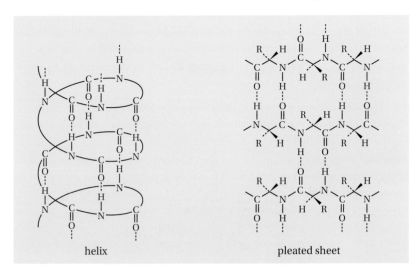

Fig 39
Hydrolysis of peptide bonds

The **primary structure** of a protein is the sequence of amino-acid units present in the polymer. To determine the primary structure, it is necessary to break down the protein systematically from one end and to identify each amino acid as it is released.

Mixtures of amino acids can be separated, and identified, by paper chromatography or by thin-layer chromatography (see section 3.4.11).

Hydrogen bonding in proteins

Peptide chains tend to form orderly, hydrogen-bonded arrangements. In particular, the carbonyl oxygen atoms form hydrogen bonds with the amide hydrogen atoms of other peptide groups, i.e. C=O---H–N. The hydrogen-bonded arrangements give rise to two distinct types of **secondary structure**, either a helix or a pleated sheet.

Despite hydrogen bonds being relatively weak, the behaviour of a large number of hydrogen bonds is cooperative, leading to the formation of stable structures. Most proteins have some helical structure. In hair and wool, the helices are coiled around each other to form rope-like structures.

helix

pleated sheet

Examiners' Notes

These structures are not required by the AQA A-level specification.

Fig 40
Types of secondary structure of proteins

3.4.9 Polymers

Essential Notes

At low temperatures, long-chain polymers are hard and glass-like. As the temperature is raised, the polymer passes through a **glass-transition temperature** (T_g) after which it becomes more flexible and mouldable (plastic); such polymers are said to be thermoplastic. In order to make synthetic polymers more flexible, non-volatile liquids (plasticisers) are added which soften the polymer and lower T_g (see section 3.4.5). For example, the value of T_g for PVC can be lowered from 80 °C to about 0 °C by the addition of dibutyl benzene-1,2-dicarboxylate.

Essential Notes

In the laboratory, nylon-6,6 can be pulled as a filament or rope from the interface between an aqueous solution of hexane-1,6-diamine and a solution of hexanedioyl dichloride in a solvent, e.g. hexane, which is immiscible with water.

Addition polymers

The polymerisation of ethene and some of its derivatives was considered in *Collins Student Support Materials: Unit 2 – Chemistry in Action*, section 3.2.9. In general, addition polymers (or **chain-growth** polymers) are made by the addition of monomers to the end of a growing chain. The end of the chain is reactive because it is a radical, a cation or an anion, according to the structure and type of catalyst used for the reaction.

Several monomers can be represented by $RCH{=}CH_2$ where R = H, CH_3, C_6H_5, Cl, CN or $OCOCH_3$. The overall reaction, an addition polymerisation, is shown below, where the **repeating unit** is enclosed in brackets; n is a variable but large number (about 100 to over 10 000).

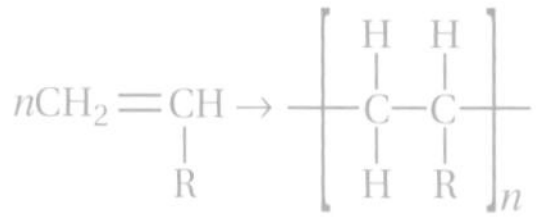

$$n CH_2{=}CH \rightarrow \left[\begin{array}{cc} H & H \\ | & | \\ C & C \\ | & | \\ H & R \end{array} \right]_n$$

It is usual to omit any consideration of end-groups, which represent an insignificant fraction of a large polymer. Some illustrative examples of addition polymers are given in Table 13.

Polyalkenes contain a skeleton consisting of carbon atoms and are therefore chemically inert, like simple alkanes, due to lack of bond polarity. Polyalkenes are non-biodegradable. Polymers of this kind are, nevertheless, highly flammable. In some cases plastics can be recycled, but the various types have to be separated from one another beforehand (see page 64).

Condensation polymers

Condensation polymers (or **step-growth** polymers) are formed by the reaction between molecules having two functional groups, involving the loss of small molecules such as H_2O, CH_3OH or HCl.

Polyamides

The reaction between a dicarboxylic acid and a diamine leads to the formation of a polyamide. Nylon-6,6 is formed from hexanedioic acid and hexane-1,6-diamine (Fig 41); the repeating unit is enclosed in brackets.

Nylon-6,6 is so called because it is derived from a six-carbon diacid and a six-carbon diamine.

Appropriate amino acids can be polymerised to form nylons. The most important example is nylon-6, which is made from 6-aminohexanoic acid, $H_2N(CH_2)_5COOH$ (Fig 42).

Aromatic polyamides (aramids) can also be made, an important example being **Kevlar**, which is derived from benzene-1,4–dicarboxylic acid and benzene-1,4–diamine:

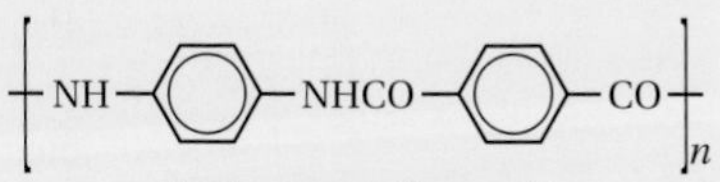

Kevlar repeating unit

Hydrogen bonding between polymer chains results in a tough, sheet-like structure:

Essential Notes

Nomex, the 1,3-linked isomer of Kevlar, is used in flame-resistant clothing for firefighters, military pilots and racing car drivers.

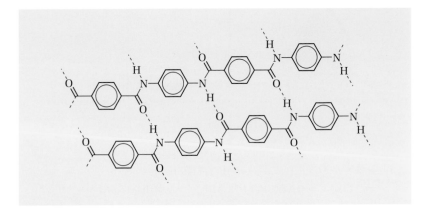

Monomer	Repeating unit	Polymer	Typical uses
$CH_2=CH_2$	$—CH_2-CH_2—$	poly(ethene) *polythene*	film, bags
$CH_2=CHCH_3$	$—CH_2-CH—$ CH_3	poly(propene) *polypropylene*	moulded plastics, fibres
$CH_2=CHC_6H_5$	$—CH_2-CH—$ C_6H_5	poly(phenylethene) *polystyrene*	packaging, insulation
$CH_2=CHCl$	$—CH_2-CH—$ Cl	poly(chloroethene) *poly(vinyl chloride)*, (PVC)	pipes, flooring
$CH_2=CHCN$	$—CH_2-CH—$ CN	poly(propenenitrile) *poly(acrylonitrile)*	fibres
$CH_2=CHOCOCH_3$	$—CH_2-CH—$ $OCOCH_3$	poly(ethenyl ethanoate) *poly(vinyl acetate)*	paints, adhesives

$CH_2=C-COOCH_3$
      $\|$
      CH_3
	$COOCH_3$
	       $\|$
	 $—CH_2-C—$
	       $\|$
	       CH_3
		poly(methyl 2-methylpropenoate)
		 poly(methyl methacrylate)
			glass replacement (Perspex), baths

$CH_2=C-COOCH_3$
      $\|$
      CN
	$COOCH_3$
	       $\|$
	 $—CH_2-C—$
	       $\|$
	       CN
		poly(methyl 2-cyanopropenoate)
		 poly(methyl cyanoacrylate)
			super-glue

$CF_2=CF_2$	$—CF_2-CF_2—$	poly(tetrafluoroethene) *poly(tetrafluoroethylene)*, (PTFE)	non-stick surfaces, non-lubricated bearings

Table 13
Some addition polymers and their typical uses

Examiners' Notes

Poly(tetrafluoroethene) has a helical structure with the fluorine atoms on the outside; the strong C—F bonds are highly resistant to chemical attack. A modern use for PTFE was in the roof of the Millennium Dome at Greenwich. The roof fabric is made up of panels of PTFE-coated fibreglass. PTFE is inert and has a low coefficient of friction, giving it non-stick properties.

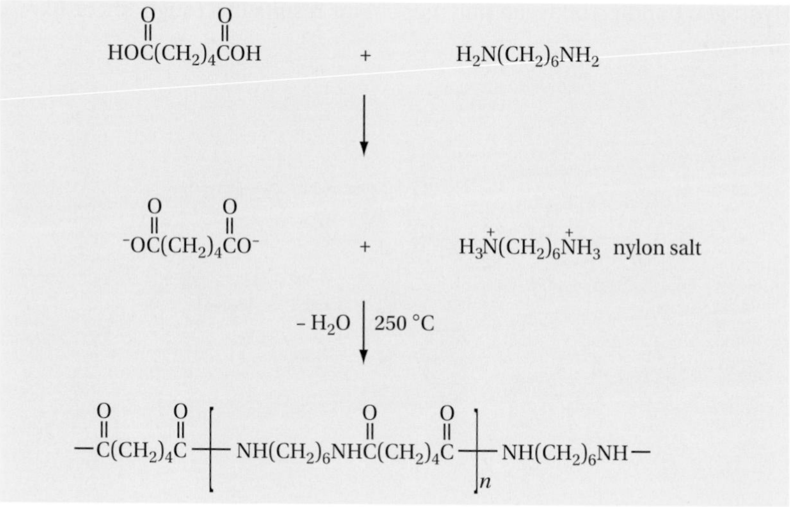

Fig 41
Formation of nylon-6,6

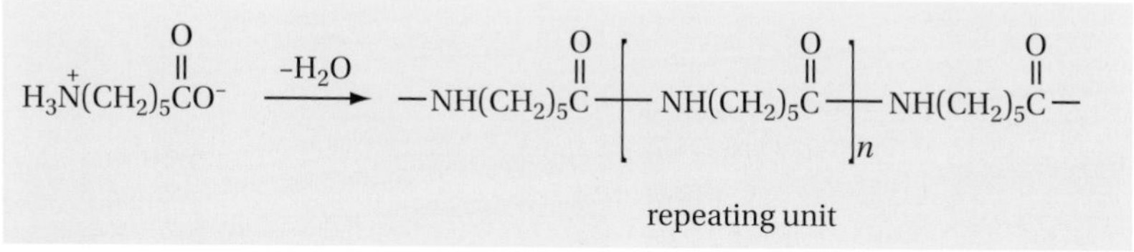

Examiners' Notes

Nylon first found widespread use in textiles and carpets. Because of its resistance to stress, other uses include tyre cords, fishing lines, mountaineering ropes and bearings and gears.

Fig 42
Formation of nylon-6

Kevlar has a tensile strength greater than that of steel and is a component of some bullet-proof vests. Other uses include the protective outer sheath of fibre optic cable and the inner lining of some bicycle tyres.

As in the case of naturally-occurring polyamides (see section 3.4.8), synthetic polyamides are susceptible to hydrolysis and can be broken down into the component monomer units. Consequently, polyamides are biodegradable polymers.

Polyesters

The reaction between a dicarboxylic acid and a diol leads to the formation of a polyester. The most important polyester is **Terylene** which can be formed from benzene-1,4-dicarboxylic (*terephthalic*) acid and ethane-1,2-diol (Fig 43).

A better product is obtained by a transesterification (see section 3.4.5) reaction between the dimethyl ester of the acid and the diol. In this case, methanol is evolved as a gas.

Polyesters, just like single esters (see section 3.4.5), are susceptible to hydrolysis and can be broken down into the component monomers. Consequently, polyesters, like polyamides, are biodegradable polymers, as shown in Fig 44.

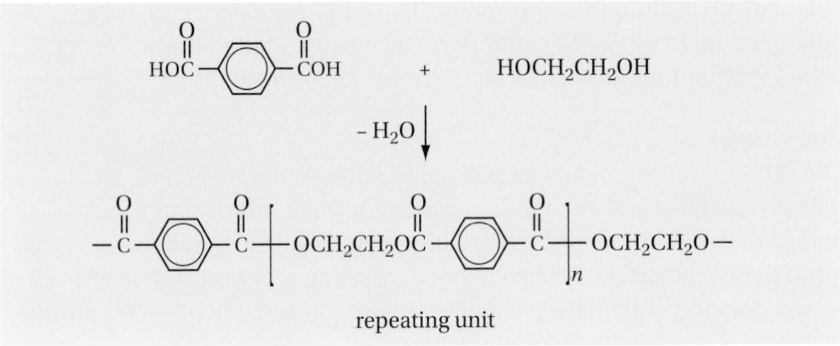

Fig 43
Formation of Terylene

Essential Notes

Molten polymers can be spun into a fibre or cast into a film. Typical uses include permanent-press fabrics and magnetic recording tapes. Terylene is also blow-moulded on a large scale to make plastic bottles.

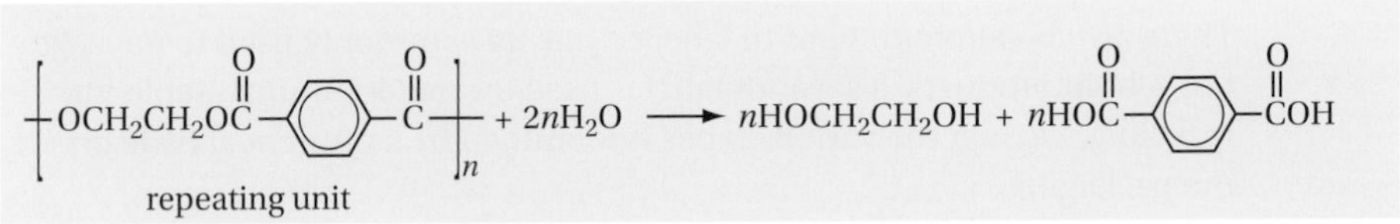

Fig 44
Hydrolysis of the polyester Terylene

Biodegradability and disposal of polymers

Most common addition polymers are not biodegradable (they cannot be broken down by micro-organisms). Polyalkenes contain no functional groups and so are chemically inert and resistant to biodegradation. Although other addition polymers contain functional groups (see Table 13), the materials they form are not biodegradable. Condensation polymers, typically polyamides and polyesters, possess repeating units which can be split by hydrolysis; such polymers are biodegradable by enzyme action.

For years, most of the plastic waste in the UK and elsewhere has gone into **landfill sites** (rubbish dumps), many of which have now reached capacity. Plastic waste accounts for only approximately 8% of the mass, but makes up about 20% of the volume of the rubbish.

The incorporation of cellulose, starch or protein into synthetic polymers or plastic articles makes them biodegradable, so that disintegration can occur in landfill sites. It is now possible to design new materials to be completely biodegradable. However, traditional synthetic plastics are currently more economically attractive than biodegradable ones. It is likely that the need for more environmentally responsible disposal will gradually lead to an increased availability of biodegradable materials.

Incineration

Polymers can be burned to recover a significant amount of energy for power generation or heating. The volume of waste is greatly reduced by such **incineration,** but the process may also generate toxic gases.

Modern incinerator designs include a high-temperature zone, where the flue gases encounter a temperature in excess of 850 °C for at least 2 seconds to ensure the complete breakdown of any organic toxins. Incineration of household waste reduces its volume by about 95%.

Essential Notes

Polymers (meaning *many units*) are effectively long chains of molecules which are usually organic or biological in nature, though some may be inorganic, e.g. silicone rubber. The term *polymer* is often used as a synonym for *plastic*. However, whereas all plastics are polymers, not all polymers are plastic. Before conversion into plastic products, polymers are often modified by the incorporation of plasticisers, stabilisers and colorants.

Examiners' Notes

Ecoflex is a fully biodegradable aliphatic–aromatic co-polyester based on butane-1,4-diol, hexanedioic acid and benzene-1,4-dicarboxylic acid. The material is used in film form for disposable packaging and will decompose within a few weeks when composted.

Before dispersal into the atmosphere, flue gases are cleaned to extract pollutants such as SO_2, NO_x and HCl. On average, about 50% of the chlorine input into municipal incinerators comes from PVC.

Recycling

The prime source of nearly all polymers is non-renewable crude oil. It makes ecological and economic sense to conserve this resource and to reduce disposal problems by **recycling** as much waste material as possible. It has to be recognised, however, that a recycling process is often far from straightforward and incurs a significant cost. Further, the resulting product may not be suitable for its original purpose.

Thermoplastics constitute the majority of disposable polymeric products. These can be melted down and reused. Successful recycling depends on accumulating enough material of a particular type to re-melt.

There are six different types of labelled plastic commonly used to package household products; a seventh label is used for materials unsuitable for recycling. Each of the various types is identified by a numerical code on the packaging:

- **Type 1: PETE** – *polyethylene terephthalate* (PET), or Terylene
  - often first used for water and beverage containers
  - recycled for use in textiles, e.g. fibres and carpets.

- **Type 2: HDPE** – *high-density polyethylene*
  - often first used for opaque milk and motor-oil containers
  - recycled for use in coloured products, e.g. bottle crates, drainage pipes and stadium seats.

- **Type 3: V** – *polyvinyl chloride* (PVC)
  - used in cling film, vegetable oil bottles and alcoholic beverage containers
  - recycled products include drainage pipes, floor tiles, non-food bottles and protective bubble wrap.

- **Type 4: LDPE** – *low-density polyethylene* [less dense and more flexible than HDPE]
  - widely used for many kinds of plastic bags, often coloured
  - recycled mainly as black garbage bags and compost bins.

- **Type 5: PP** – *polypropylene*
  - used for relatively strong food containers, e.g. ketchup bottles and margarine tubs
  - recycled products, often coloured, include cafeteria trays, ice scrapers, and VCR storage cases, as well as car bumpers and under-body parts.

- **Type 6: PS** – *polystyrene*
    - used in its expanded form for protective packing and insulated utensils
    - recycling is difficult and, as an alternative, waste material is sometimes used as a filler in other plastics and concrete.
- **Type 7: Other** – covers mixed plastics which have no recycling potential.

Great care has to be taken during recycling not to mix the various types of plastic. The presence of just one rogue PVC bottle in a melt of 10,000 PET bottles will ruin the whole batch.

At the present time, the multiple recycling of paper, glass and metals (which all become products similar to their source materials) is carried out on a much larger scale than is used for plastics, so that recycling of plastics has only a minimal impact on the amount of natural resources and energy used.

3.4.10 Organic synthesis and analysis

Applications

Synthesis

Organic chemists need to find the most efficient way to synthesise more complicated compounds from simple starting materials. For this, they rely on a sound knowledge of functional group interconversions. The Appendix contains a summary of the various organic reactions involved in *Collins Student Support Materials: Unit 2 – Chemistry in Action* and in this book.

Given a target molecule, a suitable route can often be devised in a stepwise manner by *working backwards from the product*, as well as forwards.

Example

Show how CH_3CH_2Br can be converted into $CH_3COOCH_2CH_3$.

The product, an ester, can be made from ethanoic acid and ethanol. Hydrolysis of CH_3CH_2Br will give ethanol which can, in turn, be oxidised to ethanoic acid. A likely reaction sequence is:

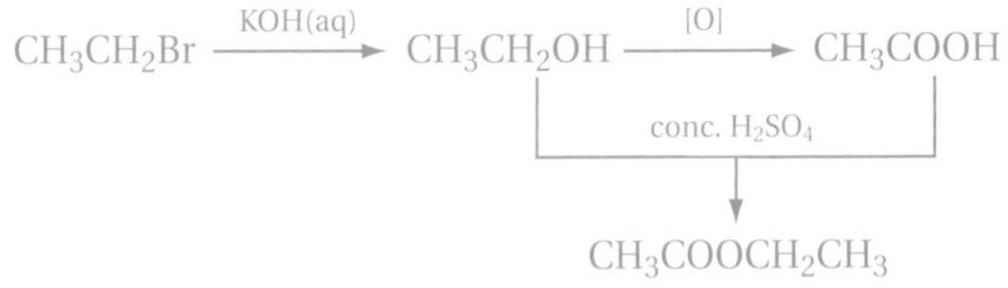

$$CH_3CH_2Br \xrightarrow{KOH(aq)} CH_3CH_2OH \xrightarrow{[O]} CH_3COOH$$

$$\xrightarrow{\text{conc. } H_2SO_4} CH_3COOCH_2CH_3$$

Example

Show how $CH_3CH=CH_2$ can be converted into $(CH_3)_2CHCH_2NH_2$.

The CH_2NH_2 section of the product suggests the reduction of a CN group as the final step.
The CN group is likely to have been introduced by a nucleophilic substitution, i.e.

$RBr \rightarrow RCN$

Addition of HBr to propene must be the first step.
The reaction sequence is therefore:

$CH_3CH=CH_2 + HBr \rightarrow$

$$CH_3\underset{Br}{CH}CH_3 \xrightarrow{KCN} CH_3\underset{CN}{CH}CH_3 \xrightarrow{Ni/H_2} CH_3\underset{CH_2NH_2}{CH}CH_3$$

Example

Outline a reaction sequence for the synthesis of $CH_3COCH_2NHCOCH_3$, starting from ethanal, using appropriate reagents.

The product contains a ketone group and a substituted amide.
Working backwards, the following conversions should come to mind:

$$CH_3COCH_2NH_2 \xrightarrow[\text{(CH}_3\text{CO)}_2\text{O}]{CH_3COCl \text{ or}} CH_3COCH_2NHCOCH_3$$

(acylation of an amine)

$$CH_3CH(OH)CH_2NH_2 \xrightarrow{[O]} CH_3COCH_2NH_2$$

(oxidation of a secondary alcohol)

$$CH_3CH(OH)CN \xrightarrow[\text{or LiAlH}_4]{H_2/Ni} CH_3CH(OH)CH_2NH_2$$

(reduction of a nitrile)

$$CH_3CHO \xrightarrow{HCN} CH_3CH(OH)CN$$

(nucleophilic addition of HCN)

The reaction sequence is therefore:

$$CH_3CHO + HCN \rightarrow CH_3CH(OH)CN \rightarrow CH_3CH(OH)CH_2NH_2 \rightarrow$$
$$CH_3COCH_2NH_2 \rightarrow CH_3COCH_2NHCOCH_3$$

Analysis

It is often necessary to identify particular functional groups by simple chemical tests. Some characteristic reactions of functional groups are shown in Table 14.

Spectroscopic methods, notably infra-red spectroscopy, are much more useful than simple chemical tests for identifying particular functional groups (see section 3.4.11).

Functional groups	Test procedure	Observation
alkenes (3.2.9) $\diagup C = C \diagdown$	add bromine water	red-brown colour decolourised
haloalkanes (3.2.8) R–X	warm with NaOH(aq); acidify with HNO_3 and add $AgNO_3$(aq)	precipitate of AgX
alcohols (3.2.10) R–OH	add acidified $K_2Cr_2O_7$	orange colour turns green with primary and secondary alcohols
	warm with CH_3COOH and a little conc. H_2SO_4	sweet smell of ester
aldehydes (3.4.5) R–CHO	add acidified $K_2Cr_2O_7$	orange colour turns green
	warm with Fehling's solution	red precipitate of Cu_2O
	warm with Tollens' reagent	silver mirror
carboxylic acids (3.4.5) R–COOH	add $NaHCO_3$(aq)	effervescence of CO_2
acyl chlorides (3.4.5) R–COCl	add $AgNO_3$(aq)	vigorous reaction and white precipitate of AgCl

Table 14
Some reactions of functional groups

3.4.11 Structure determination

Data sources

The main analytical techniques used to determine the structures of organic compounds are mass spectrometry, infra-red spectroscopy (see *Collins Student Support Materials: Unit 2 – Chemistry in Action*, section 3.2.11) and nuclear magnetic resonance spectroscopy. These techniques are often used in combination.

Mass spectrometry

The use of high-resolution mass spectrometry to determine molecular formulae was discussed in *Collins Student Support Materials: Unit 2 – Chemistry in Action*, section 3.2.11.

Fragmentation of molecular ions

The high energy of the ionising beam in the mass spectrometer causes some of the ionised molecules to break apart or **fragment**. The molecular

Essential Notes

Note that $(M - Br)^+$ or (M minus Br)$^+$ means that the ion $M^{+\bullet}$ has lost a Br atom.

ion, also called the **parent ion**, fragments to give smaller, positively charged ions, which are detected, and radicals, which are not:

$$M^{+\bullet} \to X^+ + Y^\bullet \text{ (}Y^\bullet \text{ not detected)}$$

and $M^{+\bullet} \to Y^+ + X^\bullet$ ($X^\bullet$ not detected)

A variety of fragmentation pathways is usually possible and, for each route, one of the fragments retains the positive charge and is detected. Further fragmentation of X^+ or Y^+ may also occur. The result is a characteristic relative abundance mass-spectral fragmentation pattern.

The heights of peaks on the mass spectrum are shown as *percentages of the height of the largest peak*, the **base peak**. In Fig 45, the base peak, at $m/z = 15$, is due to CH_3^+, i.e. $(M - Br)^+$. In this simple mass spectrum, peaks due to Br^+, i.e. $(M - CH_3)^+$, are easily identified for both isotopes of bromine.

Fig 45
Simplified mass spectrum of bromomethane, CH_3Br

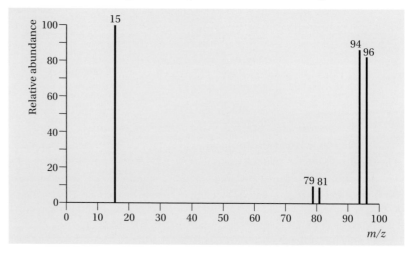

The molecular ion is sometimes shown incorrectly as just M^+. It should be $M^{+\bullet}$ to show that this species possesses an unpaired electron – it is a **radical cation** – and the fate of this unpaired electron has to be taken into account when interpreting fragmentation patterns and when writing equations for fragmentations, as seen below.

After electron impact, the molecular ions formed fragment at weaker bonds first. Dominant peaks are associated with fragments best able to bear a positive charge. However, virtually all possible pairs of fragments can be identified. The loss of methyl and ethyl radicals is very common.

Clues to molecular structure are provided when particular radicals, or sometimes molecules, are lost from the molecular ion (Table 15). Also, it is often possible to identify the formation of relatively stable ions in the mass spectrum (Table 16). The easy fragmentation of relatively weak bonds often ensures that the fragment ion $(M - X)^+$ is the base peak in the spectrum, as with bromomethane (Fig 45). In the case of alcohols, the ready elimination of water from $M^{+\bullet}$ normally gives rise to a large $(M - H_2O)^{+\bullet}$ peak 18 mass units below the parent ion.

Alkanes

Although the mass spectra of alkanes are fairly complex, due to the presence of small peaks associated with ^{13}C and with loss of hydrogen

Ion⁺	Group lost	Possible inference
$(M - 16)$	NH_2	amide
$(M - 17)$	OH	alcohol
$(M - 18)$	H_2O	alcohol
$(M - 29)$	C_2H_5	ethyl ketone
$(M - 31)$	OCH_3	methyl ester
$(M - 43)$	$COCH_3$	methyl ketone
$(M - 44)$	CO_2	carboxylic acid
$(M - 45)$	OC_2H_5	ethyl ester
$(M - 57)$	COC_2H_5	ethyl ketone
$(M - 59)$	$COOCH_3$	methyl ester
$(M - 73)$	$COOC_2H_5$	ethyl ester

Table 15
Some common groups lost from molecular ions

m/z	Ion⁺ formed
31	CH_2OH
43	CH_3CO or C_3H_7
57	CH_3CH_2CO
71	$CH_3CH_2CH_2CO$
77	C_6H_5
91	$C_6H_5CH_2$

Table 16
Some relatively stable positive ions formed from molecular ions

atoms, the more intense peaks are easily identified. In general, in unbranched alkanes the C—C bonds break more or less indiscriminately so that $(M - R)^+$ signals are found. For example, butane shows $M^{+\bullet}$ at $m/z = 58$, $(M - CH_3)^+$ at $m/z = 43$, $(M - CH_2CH_3)^+$ at $m/z = 29$ and CH_3^+ at $m/z = 15$.

The relative intensities of the various peaks differ in accordance with the stability of the carbocations formed, especially in the case of branched structures. Methylpropane, for example, shows a dominant peak at $m/z = 43$ due to C—C cleavage producing a secondary carbocation:

$$[(CH_3)_3CH]^{+\bullet} \rightarrow (CH_3)_2CH^+ + {}^\bullet CH_3$$

Ketones

The fragmentation patterns of carbonyl compounds, especially ketones, are often useful in structural identifications. The predominant decomposition pathway is at the carbonyl group to give an alkyl radical and a stable acylium cation (see Friedel–Crafts acylation in section 3.4.6):

$$[RCOR]^{+\bullet} \rightarrow RCO^+ + R^\bullet$$

An alkyl cation can also be formed, by loss of carbon monoxide from the acylium ion. There are two possible routes for fragmentation when the ketone is unsymmetrical, like butanone (Fig 46). Loss of ${}^\bullet C_2H_5$ from $M^{+\bullet}$ gives the relatively stable CH_3CO^+ at $m/z = 43$ as the base peak. Similar loss of ${}^\bullet CH_3$ from the parent ion produces the less dominant $CH_3CH_2CO^+$ at $m/z = 57$, which forms ${}^+CH_2CH_3$ when carbon monoxide is eliminated.

Essential Notes

Exercise: Identify a ketone which has a molecular ion peak at $m/z = 100$ and dominant peaks at m/z values of 71, 57, 43 and 29.

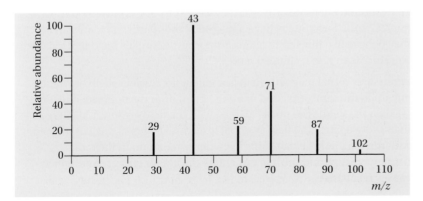

Fig 46
Simplified mass spectrum of butanone, $CH_3COCH_2CH_3$

Esters

Characteristic (M – X) peaks appear in the mass spectrum of compounds containing COOR groups, due to the loss of R, OR and COOR radicals. Other fragmentations lead to peaks associated with $^+$COOR and alkyl groups. The mass spectrum of methyl butanoate (Fig 47) is a good example.

Fig 47
Simplified mass spectrum of methyl butanoate, $CH_3CH_2CH_2COOCH_3$

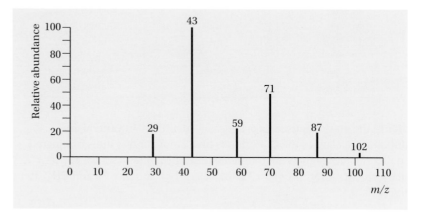

The peaks shown are due to $^+CH_2CH_3$ (29), $^+CH_2CH_2CH_3$ (43), $^+COOCH_3$ (59), $CH_3CH_2CH_2CO^+$ (71), $CH_3CH_2CH_2COO^+$ (87) and $[CH_3CH_2CH_2COOCH_3]^{+\bullet}$ (102, $M^{+\bullet}$).

Infra-red spectroscopy

The basic principles of infra-red spectroscopy were discussed in *Collins Student Support Materials: Unit 2 – Chemistry in Action*, section 3.2.11. It was shown that most functional groups in organic compounds give rise to characteristic infra-red absorptions, which change little from one compound to another.

The data provided in Table 19 of *Collins Student Support Materials: Unit 2 – Chemistry in Action*, section 3.2.11 have been augmented in this unit by the inclusion of arenes and aryl nitro derivatives (Table 17).

Illustrative infra-red absorption spectra are shown for 2-hydroxypropanenitrile (Fig 48) and propylamine (Fig 49).

Nuclear magnetic resonance spectroscopy

Many atomic nuclei behave as if they were spinning and are said to have **nuclear spin**. Such nuclei have a **magnetic moment,** which means that they behave like tiny bar magnets. Important examples include ^1H (called a proton in an n.m.r. context), and ^{13}C.

When an external magnetic field is applied, the nuclei, which have spin, line up in the same direction (with the field) or in the opposite direction (against the field). The nuclei aligned with the field are lower in energy than those against the field. A signal is recorded when a nucleus aligned with the magnetic field absorbs low-energy radiation in the radio-frequency range and the nucleus enters the higher energy state, which causes **resonance**.

Essential Notes

Nuclei possessing even numbers of both protons and neutrons, such as ^{12}C or ^{16}O, lack magnetic properties and do not give rise to n.m.r. signals.

Bond	Types of compound	Range/cm^{-1}
C–H	alkanes	2850–2960
	alkenes	3010–3095
	alkynes	3250–3300
	arenes	3030–3080
	aldehydes	2710–2730
O–H	alcohols (H-bonded)	3230–3550
	carboxylic acids (H-bonded)	2500–3000
N–H	amines	3320–3560
C–C	alkanes	750–1100
C=C	alkenes	1620–1680
C≡C	alkynes	2100–2250
C⋯C	arenes	1500–1600
C–O	alcohols, ethers, carboxylic acids, esters	1000–1300
C=O	aldehydes, ketones, carboxylic acids, esters	1680–1750
C–N	amines	1180–1360
C≡N	nitriles	2210–2260
C–Cl	haloalkanes	600–800
C–Br	haloalkanes	500–600
N–O	aryl nitro derivatives	1330–1550

Table 17
Some characteristic infra-red absorptions due to bond stretching in organic molecules

Essential Notes

Note that appropriate infra-red data will be provided in examination questions.

Examiners' Notes

Absorptions due to various kinds of C—H bending in saturated groups appear characteristically at about 1460 cm^{-1} and 1370 cm^{-1}.

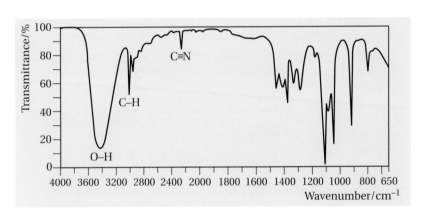

Fig 48
Infra-red spectrum of 2-hydroxypropanenitrile, $CH_3CH(OH)CN$

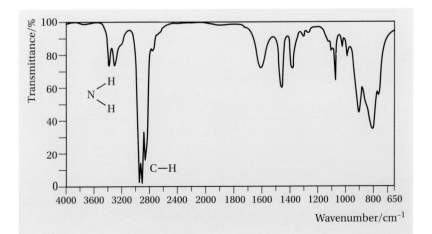

Fig 49
Infra-red spectrum of propylamine, $CH_3CH_2CH_2NH_2$

Essential Notes

In the case of NH_2 groups, two peaks are usually seen in the same region, due to both symmetric and asymmetric vibrations.

Examiners' Notes

Note that the word 'proton' as used in n.m.r. spectroscopy is not an H^+ ion, but refers to a magnetically responsive hydrogen atom – whose nucleus is a proton.

Examiners' Notes

Examination questions involving n.m.r. will not require a knowledge of n.m.r. theory. Candidates will only need to apply their understanding of n.m.r. spectra to problem solving..

^1H n.m.r. spectroscopy

Protons in organic molecules

In organic molecules, protons are surrounded by electrons which partly shield them from the applied magnetic field. The amount of shielding depends on the electron density surrounding the nucleus and varies for different protons within a compound. Factors which influence the electron density include bond polarity and the presence of electron-donating or electron-withdrawing groups:

● The nucleus is **deshielded** when the electron density is reduced.

● The nucleus is **shielded** when the electron density is increased.

Because each chemically distinct hydrogen atom (proton) has a unique electronic environment, it gives rise to a characteristic resonance. *Chemically equivalent protons* are all in the same environment and therefore absorb at the same frequency (see page 73). For hydrogen atoms, the differences in frequency are tiny, being recorded as only a few **parts per million** (ppm).

Chemical shift

The movements caused by shielding (moving the recorded value *upfield*) and by deshielding (moving the recorded value *downfield*) are quantified by the **chemical shift**, δ. Chemical shifts are measured, in ppm, relative to an internal standard, tetramethylsilane (TMS), $(CH_3)_4Si$, because:

● TMS gives a signal that resonates upfield from almost all other organic hydrogen resonances, because the 12 equivalent hydrogens are highly shielded.

● TMS gives a single, intense peak since there are 12 equivalent protons.

● TMS is non-toxic and inert.

● TMS has a low boiling point (26.5 °C) and can easily be removed from a sample.

By definition, the δ value of $(CH_3)_4Si$ is zero. Almost all proton n.m.r. absorptions occur 0–10 δ downfield from $(CH_3)_4Si$ (Fig 50). The δ values vary according to the *structural* environment, so that organic functional groups have characteristic chemical shifts.

Fig 50
The δ scale of chemical shifts

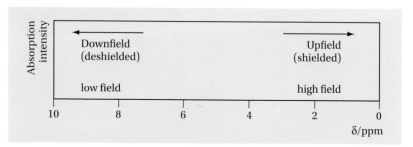

Proton n.m.r. spectra are recorded in solution. The sample to be examined (a few milligrams) is dissolved in a proton-free solvent to avoid unwanted absorptions. Typical solvents include CCl_4 and deuterated compounds, such as $CDCl_3$ and C_6D_6.

Features of proton n.m.r. spectra

There are four important aspects of ^1H n.m.r. spectra which provide information about chemical structure:

- The *number of absorptions* indicates how many kinds of non-equivalent protons are present.
- The *intensities of the absorptions* reveal how many protons are associated with each resonance peak.
- The *positions of the absorptions* give clues as to the environment of each kind of proton.
- The *splitting of an absorption* into several peaks, which is caused by *spin–spin coupling*, provides information about neighbouring protons.

High-resolution proton n.m.r. spectroscopy has become a very powerful tool for elucidating organic chemical structures.

Number of absorptions

The different environments of the protons in an organic molecule are revealed by an examination of its low-resolution ^1H n.m.r. spectrum. In the case of ethanol, for example, three distinct signals are found (Fig 51).

Essential Notes

High-resolution spectroscopy is able to reveal details of the structure of a peak that are obscured in low-resolution spectroscopy.

Fig 51
Low-resolution ^1H n.m.r. spectrum of ethanol, CH_3CH_2OH

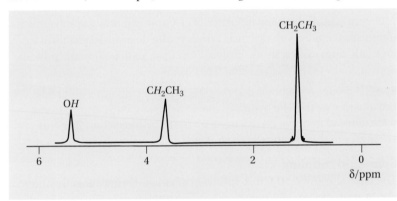

The three kinds of non-equivalent proton are seen at different positions and the peaks are of different intensities. For each peak, the area is proportional to the number of hydrogen atoms giving rise to the signal. Thus, for ethanol, the areas under the peaks are in the ratio of 1:2:3, in accordance with $OH:CH_2:CH_3$. The spectrometer is able to measure the **relative intensities** of the various peaks electronically and provide an **integration trace**, which calculates the area under each peak. Integrated peak areas are sometimes presented on the recorded spectrum in a stepwise manner, so that the height of each step is proportional to the number of hydrogen atoms associated with each signal (see Fig 53).

Position of absorption

As pointed out earlier, the effects of shielding and deshielding are expressed by chemical shift values on the δ scale. For ethanol (Fig 51), the hydroxyl hydrogen atom is deshielded because of the effect of the electron-withdrawing electronegative oxygen atom. However, the CH_2 and CH_3 groups are progressively further from the oxygen atom and the electron density around the hydrogen atoms increases. The CH_3 hydrogen atoms are the most highly shielded and absorb at the higher field.

In general, the positions of absorption (as measured by chemical shifts) can be related to the electronegativities of adjacent atoms and the electron-withdrawing or electron-donating effects of functional groups (Table 18).

Table 18
Some typical proton chemical shifts

Type of proton	δ/ppm
RCH_3	0.8–1.0
R_2CH_2	1.2–1.4
R_3CH	1.4–1.6
$RCOCH_3$	2.1–2.6
RCH_2OR	3.3–3.9
RCH_2Br	3.4–3.6
RCH_2Cl	3.6–3.8
$R_2C=CH_2$	4.6–5.0
ArH	6.0–8.5
RCHO	9.5–9.9

Examiners' Notes

Note that the δ values in Table 18 are only approximate because they are affected by neighbouring substituents. Appropriate data will be provided, as necessary, for use in examination questions.

Multiple substitution has a cumulative effect. Thus, for example, the δ values for CH_3Cl, CH_2Cl_2 and $CHCl_3$ are, respectively, 3.05, 5.30 and 7.27 ppm. The deshielding influence of electron-withdrawing substituents diminishes rapidly with distance, however. The delocalised electrons in aromatic rings exert a strong deshielding effect, so that aromatic protons appear at low field – high δ value (see Fig 55).

The OH groups of alcohols and the NH_2 groups of amines exhibit relatively broad n.m.r. peaks. The hydrogen atoms in these groups absorb over a wide range of frequencies ($\delta = 0.5$–6.0) due to hydrogen bonding and sensitivity to the solvent and to moisture.

Spin–spin coupling

The high-resolution ^1H n.m.r. spectrum of ethanol reveals that the three absorptions are not all single peaks, called **singlets**, (Fig 52): the CH_3 absorption signal is split into three peaks, a **triplet**, whereas the CH_2 signal appears as four peaks, a **quartet**. This situation arises because non-equivalent hydrogens on adjacent atoms interact – *couple* – with one another.

Examiners' Notes

Note that equivalent hydrogens on adjacent atoms do not display any coupling effects.

In general, **splitting** of single absorption peaks into more complex patterns of so-called **multiplets** is due to coupling between neighbouring nuclear spins. Thus, the spin of one proton can couple with the spins of adjacent protons. Splitting is observed only between nuclei with *different* chemical shifts.

The splitting of an absorption signal is described by the ***n*** + **1 rule**:

Definition

Signals for protons adjacent to n equivalent neighbours split into n + 1 peaks.

For a molecule such as $ClCH_2CHCl_2$, the two n.m.r. signals (from the CH_2 and CH protons) are split into a **doublet** and a **triplet**, respectively (Fig 53).

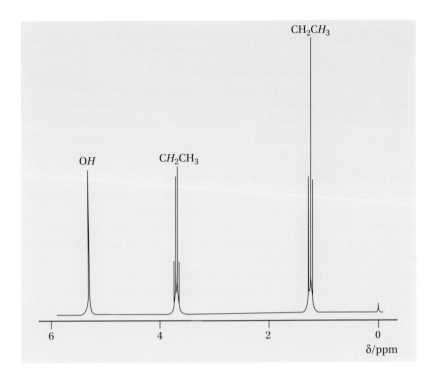

Fig 52
High-resolution ^{1}H n.m.r. spectrum of ethanol, CH_3CH_2OH

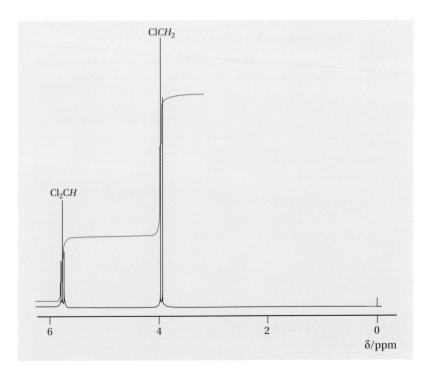

Fig 53
^{1}H n.m.r. spectrum of 1,1,2-trichloroethane, $ClCH_2CHCl_2$

Examiners' Notes

The multiplicity and relative intensities of the $n + 1$ components are given by the coefficients of the terms in the expansion of $(1 + x)^n$. These values can be obtained from Pascal's triangle

$$
\begin{array}{ccccccccccccc}
 & & & & & & 1 & & & & & & \\
 & & & & & 1 & & 1 & & & & & \\
 & & & & 1 & & 2 & & 1 & & & & \\
 & & & 1 & & 3 & & 3 & & 1 & & & \\
 & & 1 & & 4 & & 6 & & 4 & & 1 & & \\
 & 1 & & 5 & & 10 & & 10 & & 5 & & 1 & \\
1 & & 6 & & 15 & & 20 & & 15 & & 6 & & 1 \\
\end{array}
$$

where each inner number is the sum of the two numbers closest to it in the row above.

The peak for the two equivalent CH_2 protons is split into a doublet $(1 + 1)$ by the single adjacent proton. The peak for the CH proton, however, is split into a triplet $(2 + 1)$ by the two CH_2 protons.

In the case of ethanol (Fig 52), the CH_3 protons appear as a triplet (2 + 1) because of spin–spin coupling with the CH_2 protons. The CH_2 group is split into a quartet (3 + 1) by the protons of the methyl group.

The OH absorption is seen as a single peak, devoid of any splitting. Since the hydroxyl group is next to a methylene group, spin–spin coupling should result in a triplet. However, the weakly acidic OH hydrogens exchange rapidly between other ethanol molecules and also water molecules, normally present in trace amounts (chemical exchange). Absorptions of this type are said to be *decoupled* by fast proton exchange.

Interpretation of proton ^1H n.m.r. spectra

Most ^1H n.m.r. spectra exhibit splitting patterns. Equivalent nuclei located next to *one* neighbouring hydrogen atom resonate as a 1:1 *doublet* (d). Equivalent nuclei located next to a group of *two* equivalent hydrogens resonate as a 1:2:1 *triplet* (t). Equivalent nuclei adjacent to a set of *three* equivalent hydrogens resonate as a 1:3:3:1 *quartet* (q). Such patterns are often encountered in simple organic structures. The ^1H n.m.r. spectra of ethyl ethanoate (Fig 54) and 2-phenylethyl ethanoate (Fig 55) illustrate these principles at work.

^{13}C n.m.r. spectroscopy

^{13}C n.m.r. spectroscopy provides direct information about the carbon skeleton of a molecule, and it is possible to determine:

- The *number* of non-equivalent carbon atoms in the structure (i.e. the *number* of carbon atoms in different chemical environments).

- The *different types* of carbon atom present in the compound (typically saturated, unsaturated, aromatic and carbonyl carbon atoms).

Examiners' Notes

Questions about splitting in AQA examinations will be restricted to a consideration of protons that have 1, 2 and 3 neighbouring hydrogen atoms only, giving rise to doublets, triplets and quartets, respectively.

Fig 54
^1H n.m.r. spectrum of ethyl ethanoate,
$CH_3COOCH_2CH_3$

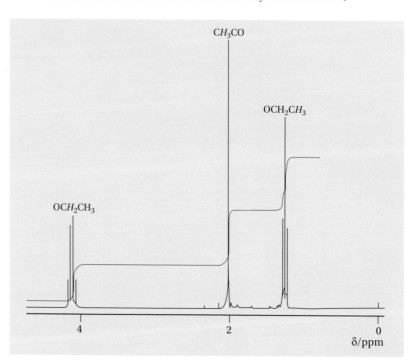

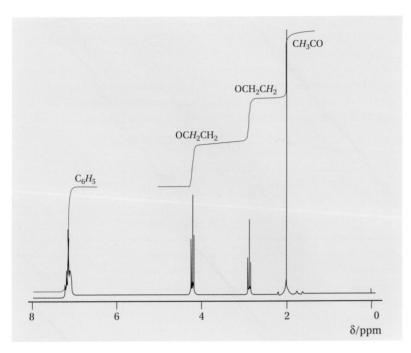

Fig 55
^{1}H n.m.r. spectrum of 2-phenylethyl ethanoate, $CH_3COOCH_2CH_2C_6H_5$

Examiners' Notes

Examination questions involving the spin–spin coupling of aromatic protons will not be set.

The interpretation of ^{13}C n.m.r. spectra is, in general, easier than in the case of ^{1}H n.m.r. spectra. However, both techniques are often used in combination in structure elucidation, usually together with infra-red and mass spectral data.

The ^{13}C isotope is magnetically active, in the same way as ^{1}H, but its natural abundance is only 1.1%. The resonances of ^{13}C nuclei are therefore more difficult to observe and are much weaker than proton resonances. Consequently, a ^{13}C n.m.r. spectrum has to be built up from a collection of molecules, since an individual molecule is unlikely to contain more than one ^{13}C nucleus and is therefore unable to provide more than a single ^{13}C resonance. Thus, a greater number of individual scans of the spectrum must be accumulated than is the case for a proton spectrum.

As for ^{1}H spectra, the **chemical shift** (δ) is an important parameter. By definition, the δ value of the internal standard tetramethylsilane, $(CH_3)_4Si$, is zero, as it is for proton n.m.r. Here, however, it is the methyl group *carbon* atoms that are used for reference, as opposed to the methyl group *hydrogen* atoms. Each non-equivalent carbon atom in an organic molecule gives rise to a signal with a quite different ^{13}C chemical shift; the typical range of observed chemical shifts (0 to 220 ppm) in ^{13}C n.m.r. spectroscopy is much larger than that for protons (0 to 12 ppm). Because of this wide range of values, and in contrast to many proton n.m.r. spectra, the different ^{13}C peaks are less likely to overlap. Note that, in contrast to ^{1}H n.m.r. spectroscopy, the factors affecting ^{13}C shifts operate, in the main, through only one bond.

In nearly the same way as for ^{1}H spectra, the positions of ^{13}C absorptions can be related to the electronegativities of adjacent atoms and the electron-withdrawing or electron-donating effects of functional groups. Consequently, it is found that, for a particular organic compound, the signals arising from the non-equivalent carbon atoms present tend to follow the same relative order as those from the protons on those carbon atoms. In saturated molecules, the downfield shift brought about by the deshielding effect of an electronegative element is greater for a ^{13}C atom than it is for a proton, because this effect occurs through only a single bond (X—C, i.e. from X to ^{13}C); in the case of protons, the effect is transmitted through two bonds (X—C—H, i.e. from X via C to ^{1}H).

Some typical ^{13}C chemical shift ranges are given in Table 19.

Table 19
Some typical ^{13}C chemical shift ranges

Type of carbon	δ/ppm
R$\underline{C}$H$_3$	5–30
R$_2\underline{C}$H$_2$	15–40
R$_3\underline{C}$H	20–40
R$_4\underline{C}$	25–40
$\underline{C}$N (amine)	25–60
$\underline{C}$C=O	20–50
$\underline{C}$Br	10–60
$\underline{C}$Cl	25–70
$\underline{C}$OR (alcohol, ether)	50–75
O=CO$\underline{C}$ (ester)	50–80
$\underline{C}$N (nitrile)	110–125
$\underline{C}$=$\underline{C}$ (alkene)	100–150
$\underline{C}$=$\underline{C}$ (aromatic)	110–160
$\underline{C}$=O (ester, amide, acid)	160–185
$\underline{C}$=O (aldehyde, ketone)	190–220

Due to the low natural abundance of ^{13}C, spin–spin coupling between two adjacent ^{13}C atoms in the same molecule is extremely unlikely and can be ignored; the chance of finding more than one ^{13}C atom in a molecule is almost nil.

Although spin–spin coupling between ^{13}C and ^{1}H does occur, most ^{13}C n.m.r. spectra are obtained as **proton-decoupled spectra**, in which only singlets are observed for each of the non-equivalent carbon atoms present. This instrumental technique greatly simplifies the spectrum and avoids overlapping multiplets but, as a consequence, all the information on the number of attached hydrogen atoms is lost. At the same time, the intensities of some of the carbon resonances increase beyond those observed in the corresponding proton-coupled spectrum; such enhancement increases, but not always linearly, with the number of hydrogen atoms attached. This enhancement is called the nuclear Overhauser effect and is named after its discoverer.

Integral information derived from ^{13}C n.m.r. spectra is not directly proportional to the number of atoms giving rise to the signal. However, a singlet peak derived from two (equivalent) carbon atoms is larger than one derived from a single carbon atom. It is often, but not always, found that a CH$_3$ peak has a greater intensity than a CH$_2$ peak, which in turn is more intense than a CH peak; quaternary carbon atoms are normally the weakest peaks in the spectrum.

Examiners' Notes

You are **not** required to know about proton-coupled spectra. Examination questions will only refer to proton-decoupled spectra, with no splitting of the ^{13}C peaks.

The proton-decoupled ^{13}C n.m.r. spectrum of methyl ethanoate is shown in Fig 56; the assignments given are consistent with the values provided in the chemical shift table (Table 19). Thus, the methyl carbon atom (*c*), adjacent to the ester carbonyl group, is found at high field (δ = 20.6 ppm), whereas the methyl carbon atom (*b*), next to the electronegative oxygen atom, is further downfield (δ = 51.6 ppm); the ester carbonyl group (*a*) appears as a singlet at low field (δ = 171.5 ppm).

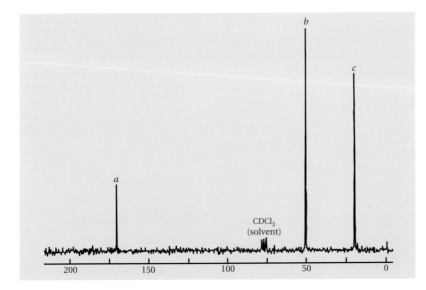

Fig 56
Proton-decoupled ^{13}C n.m.r. spectrum of methyl ethanoate

CH_3COOCH_3
c   *a*   *b*

a = 171.5 ppm
b = 51.6 ppm
c = 20.6 ppm

Comparisons between the ^{1}H and the ^{13}C n.m.r. spectra of the same compound can be useful and instructive. Ethyl ethanoate provides a simple example. The ^{1}H n.m.r. spectrum of this ester appears in Fig 54, revealing three non-equivalent groups of hydrogen atoms. The corresponding proton-decoupled ^{13}C n.m.r. spectrum (Fig 57) shows the presence of four non-equivalent carbon atoms in the structure; the signal at approximately 77 ppm is that of the solvent ($CDCl_3$).

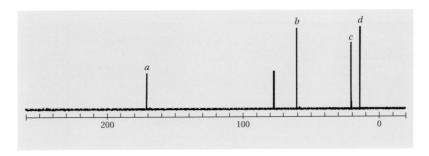

Fig 57
Proton-decoupled ^{13}C n.m.r. spectrum of ethyl ethanoate

$CH_3COOCH_2CH_3$
c   *a*   *b* *d*

a = 171.1 ppm
b = 60.4 ppm
c = 21.0 ppm
d = 14.3 ppm

The relative simplicity of ^{13}C n.m.r. spectra is useful in the identification of increasingly large alkyl groups, where complex multiplets are observed in the corresponding ^{1}H n.m.r. spectra. For example, the ^{1}H n.m.r. spectrum of butyl ethanoate is shown in Fig 58. There are five non-equivalent groups of protons present in this ester, giving rise to a singlet (δ = 1.87 ppm), two triplets (δ = 3.90 ppm and δ = 0.78 ppm) and two unresolved multiplets (which are centred at δ = 1.45 ppm and δ = 1.23 ppm).

Fig 58
^{1}H n.m.r. spectrum of butyl
ethanoate

Fig 58
^{1}H n.m.r. spectrum of butyl
ethanoate

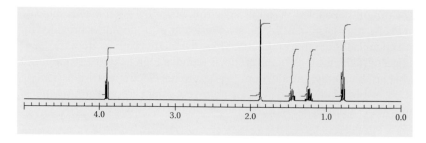

The proton-decoupled ^{13}C n.m.r. spectrum of the same ester (Fig 59)
reveals the presence of six non-equivalent carbon atoms. This kind of
improved resolution is possible with molecules of much greater
complexity.

Fig 59
Proton-decoupled ^{13}C n.m.r.
spectrum of butyl ethanoate

$CH_3COOCH_2CH_2CH_2CH_3$
$d \quad a \quad\quad b \quad c \quad e \quad f$

$a = 170.9$ ppm
$b = 64.1$ ppm
$c = 30.6$ ppm
$d = 20.8$ ppm
$e = 19.0$ ppm
$f = 13.5$ ppm

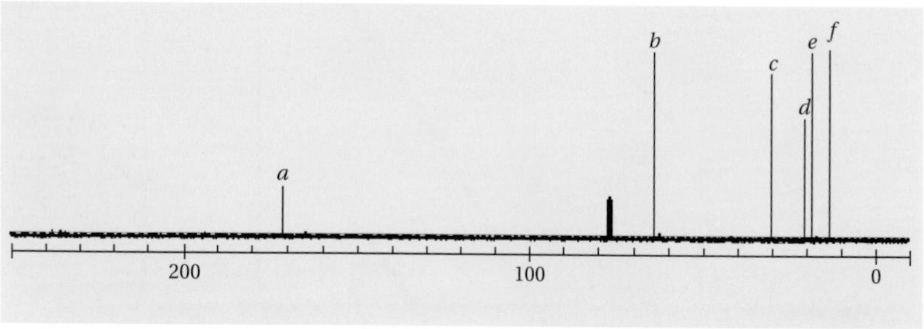

^{13}C n.m.r. spectroscopy is especially useful in the analysis of substitution
patterns in benzene rings and for the identification of isomers. Most
polysubstituted benzene rings display six different peaks in the proton-
decoupled spectrum, one for each carbon atom; symmetrical substitution
can reduce this number. Dichlorobenzene provides a straightforward
example. Symmetry considerations reveal that 1,2-dichlorobenzene has
three unique carbon atoms (3 peaks), 1,3-dichlorobenzene has four non-
equivalent carbon atoms (4 peaks) whereas 1,4-dichlorobenzene has only
two such atoms (2 peaks):

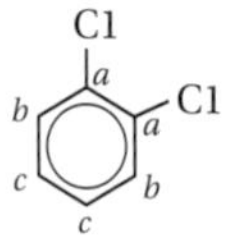

1,2-
3 peaks

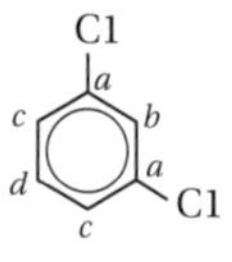

1,3-
4 peaks

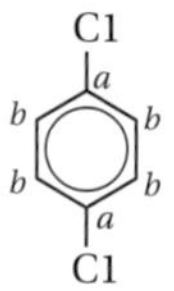

1,4-
2 peaks

Chromatography

Chromatography is the collective term used for various related laboratory
techniques that permit the separation and identification of the chemical
components in a mixture. All forms of chromatography involve a fixed
stationary phase through which passes a **mobile** or **moving phase**
containing the mixture to be separated. Separation is achieved because

components of the mixture distribute themselves differently between the two phases according to their affinity for each phase.

> **Definition**
> *Chromatography is a technique for separating the components of a mixture on the basis of differences in their affinities for a stationary and for a moving phase.*

It is helpful to make a distinction between **analytical** chromatography and **preparative** chromatography. Analytical chromatography operates with small amounts of material and aims to identify and measure the relative proportions of the various components present in a mixture; such an approach often involves comparisons with known standards for purposes of identification. Preparative chromatography is carried out on a relatively large scale and is a form of purification.

Two main kinds of chromatography can be distinguished:

- **partition** chromatography
- **adsorption** chromatography.

In partition chromatography, the stationary phase is a thin, non-volatile liquid film held on the surface of an inert solid or within the fibres of a supporting matrix; the moving phase is a liquid or a gas. Solute molecules equilibrate, or **partition**, between the two phases. Separation depends on the balance between solute solubility in the moving phase and retention in the stationary phase.

In adsorption chromatography, the stationary phase is a solid, such as alumina (Al_2O_3), and the moving phase is a liquid or a gas. The surface area of the solid phase is maximised by the use of finely-divided particles. Solute molecules become attached to **adsorption** sites on the stationary phase. Strongly adsorbed molecules travel more slowly in the moving phase than those that are only weakly adsorbed.

Types of chromatography included within the two main categories are:

- **paper** chromatography
- **thin-layer** chromatography (TLC)
- **column** chromatography
- **gas–liquid** chromatography (GLC).

Paper chromatography

This technique can be regarded as an example of partition chromatography where the stationary phase is essentially a thin layer of water adsorbed on the cellulose fibres of the paper. The moving phase is a solvent or a solvent mixture.

Drops of concentrated sample solution are spotted along a baseline, drawn in pencil near the bottom of a piece of chromatography paper. The dried paper is then dipped in a shallow layer of solvent inside a suitable container and sealed (see Fig 60). The solvent rises up the paper, carrying along the solute components; these travel at different rates according to

> **Examiners' Notes**
> The term *chromatography* is coined from two Greek words, χρωμα (chroma), meaning colour, and γραφειν (graphein), meaning to write. The word was first used in 1906 by the Russian botanist Mikhail Tsvet, who invented the technique in 1901 when he separated chlorophyll and carotenoid plant pigments down a column of calcium carbonate.

> **Examiners' Notes**
> Paper chromatography is an effective method for revealing the coloured components present in some sweets (typically Smarties) and in ballpoint pen inks or felt-tip markers.

Fig 60
Paper chromatography of three
simple mixtures

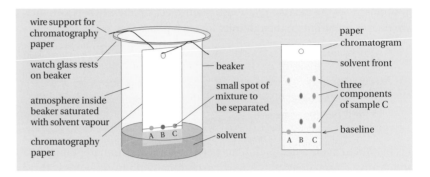

their different solubilities in the moving and stationary phases, resulting in separation. The process is stopped when the leading edge of the solvent, the **solvent front**, gets close to the top of the paper. The resulting **chromatogram** is taken out and dried.

> **Definition**
>
> A **chromatogram** is a pattern of separated substances obtained by chromatography.

The ratio of the distance travelled up the paper by a component relative to that of the solvent is called the **retention factor, R_f**:

$$R_f = \frac{\text{distance travelled by the compound}}{\text{distance travelled by the solvent front}}$$

Under standard conditions, R_f values can be used as a means of identification. The most reliable approach is to compare both known and unknown compounds together on the same chromatogram.

When, as in the majority of cases, the constituents of a mixture are colourless, it is necessary to treat the chromatogram to make the components visible. Heating the chromatogram will sometimes suffice, as will exposure to ultra-violet light.

In some cases, the components of a mixture cannot be **resolved**, i.e. separated completely, by paper chromatography in one direction. The problem can often be overcome by repeating the process using a different solvent, after rotating the paper through 90°, in order to produce a two-dimensional array of components (see Fig 61).

Fig 61
Two-dimensional paper
chromatography

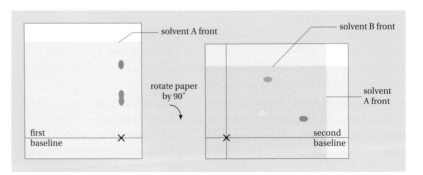

Thin-layer chromatography (TLC)

This laboratory technique is similar to paper chromatography. The stationary adsorbent phase is a thin layer of a polar matrix, such as silica gel or alumina, coated as a paste on to a glass plate or on to a thick aluminium or plastic sheet and then dried. A concentrated solution of the mixture is applied as a spot or a streak near the bottom of the plate. The treated plate is dipped in a shallow layer of solvent (the moving phase) which gradually rises up the plate by capillary action, leading to separation of the components; solutes travel different distances according to how strongly they interact with the adsorbent.

The plate is taken out and dried when the solvent front has almost reached the top. The chromatogram can be used to provide R_f values in the same way as for paper chromatography.

Visualisation of the spots on a chromatogram is often helped by the deliberate addition of a fluorescent inorganic compound to the adsorbent. The coated plate displays an overall pale green fluorescence under ultra-violet exposure; this fluorescence is quenched by the chemicals present in the spots, which renders them visible.

TLC has several advantages over paper chromatography:

- the procedure is quicker
- separations are more efficient
- results are more easily reproduced
- the adsorbent can be varied.

Thin-layer adsorption chromatography is used widely to monitor the course of organic reactions. It is often possible to follow the gradual disappearance of starting material and the appearance of product.

Column chromatography

This simple technique is an example of adsorption chromatography where the stationary phase is finely-divided alumina or silica gel contained in a vertical glass tube (the column). The moving phase (called the **eluent**) is usually an organic solvent.

A solution of the mixture is added to the top of the column, followed by enough fresh solvent to wash the components down the column; this process is referred to as **elution**.

The most strongly adsorbed components take the longest time to flow through the column. In general, the more polar the molecule, the greater the **retention time**. The eluent is either a pure solvent or a mixture of solvents chosen so that the different compounds can be separated effectively. Ideally, the components should elute one at a time from the column. Coloured components can be seen through the glass wall of the column as moving bands; fluorescent compounds can be visualised with the aid of an ultra-violet lamp. The solution emerging from the column (the **eluate**) is normally collected throughout the chromatographic separation as a series of fractions. Each fraction can be inspected separately and analysed for dissolved compounds.

Examiners' Notes

Two-dimensional paper chromatography is quite powerful and was used by the biochemist Frederick Sanger in 1955 to identify the 17 amino acids generated by the acid hydrolysis of the protein insulin. Amino-acid spots on a chromatogram are made visible as bluish-purple stains by treatment with a solution of the reagent ninhydrin.

Examiners' Notes

Thick-layer adsorption chromatography can be used as a purification technique for up to 100 mg of a sample. The purified, adsorbed product can be scraped off the plate, dissolved in a suitable solvent and recovered.

Essential Notes

In a school laboratory, it is sometimes convenient to use an ordinary burette as a chromatography column.

> **Definition**
>
> *The **retention time** is the time each component remains in the column.*

Column chromatography provides a convenient way of separating and purifying individual organic compounds from mixtures. The whole process can be speeded up by forcing the solvent through the system under pressure (flash chromatography).

Gas–liquid chromatography (GLC)

This analytical technique is a powerful method for the separation of mixtures of volatile compounds. The procedure uses a **carrier gas** (acting as eluent), such as helium or nitrogen, as the moving phase, and an inert powder coated with a film of a non-volatile liquid as the stationary phase. This coated powder is packed into a long, narrow-bore stainless steel or glass coiled tube (the column). Some chromatographs use very long, coiled capillary columns coated on the inside with the active film of non-volatile liquid.

> **Definition**
>
> *A **chromatograph** is the apparatus used for chromatographic separation of volatile components in a mixture.*

A continuous, steady flow of carrier gas passes through the column; the higher the flow rate the faster the analysis, but the lower the resolution.

The vaporised sample is injected into the entrance (head) of the column and the components are carried through the system and appear later, in sequence, at the exit. Each component has a characteristic retention time that depends on factors such as the nature of the stationary phase, the operating temperature, the flow rate of the carrier gas and the length of the column. Solubility in the non-volatile liquid phase is of major importance. A component that is highly soluble in the liquid phase takes considerably longer to elute from the column than one having a low solubility.

A detector is used to monitor the outlet stream from the column and this is linked to a recorder, so that each component appears as a peak on a chart. In conjunction with internal standards, retention times can be used to identify components, and peak areas are proportional to the amounts of different substances present.

GLC can be used to measure minute quantities of chemicals and to distinguish between closely related groups of molecules. Very often, the chromatograph is connected to a mass spectrometer which operates as a detector capable of analysing in detail each separated component present in the mixture as it emerges from the column.

Appendix

Summary of organic reactions in Units 1, 2 and 4

Alkanes

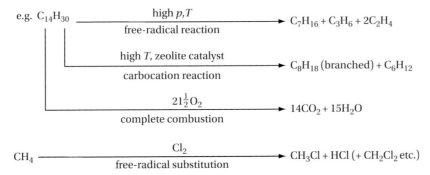

e.g. $C_{14}H_{30}$

$$\xrightarrow[\text{free-radical reaction}]{\text{high } p, T} C_7H_{16} + C_3H_6 + 2C_2H_4$$

$$\xrightarrow[\text{carbocation reaction}]{\text{high } T, \text{ zeolite catalyst}} C_8H_{18} \text{ (branched)} + C_6H_{12}$$

$$\xrightarrow[\text{complete combustion}]{21\frac{1}{2}O_2} 14CO_2 + 15H_2O$$

$$CH_4 \xrightarrow[\text{free-radical substitution}]{Cl_2} CH_3Cl + HCl \ (+ CH_2Cl_2 \text{ etc.})$$

Alkenes

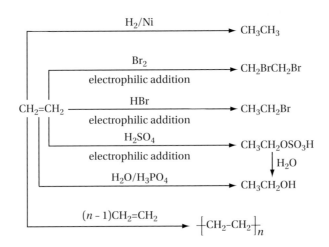

$$\xrightarrow{H_2/Ni} CH_3CH_3$$

$$\xrightarrow[\text{electrophilic addition}]{Br_2} CH_2BrCH_2Br$$

$$CH_2=CH_2 \xrightarrow[\text{electrophilic addition}]{HBr} CH_3CH_2Br$$

$$\xrightarrow[\text{electrophilic addition}]{H_2SO_4} CH_3CH_2OSO_3H \xrightarrow{H_2O}$$

$$\xrightarrow{H_2O/H_3PO_4} CH_3CH_2OH$$

$$\xrightarrow{(n-1)CH_2=CH_2} \left[CH_2-CH_2\right]_n$$

$$CH_3CH=CH_2 \xrightarrow[\text{(2) } H_2O]{\text{(1) } H_2SO_4} CH_3CH(OH)CH_3 \text{ major product}$$

$$\xrightarrow{HBr} CH_3CHBrCH_3 \text{ major product}$$

$$\xrightarrow{(n-1)CH_3CH=CH_2} \left[\begin{array}{c} CH-CH_2 \\ | \\ CH_3 \end{array}\right]_n$$

Haloalkanes

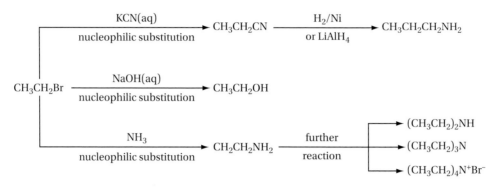

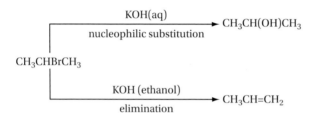

Alcohols

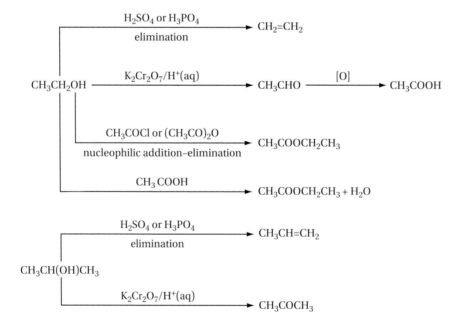

Aldehydes and ketones

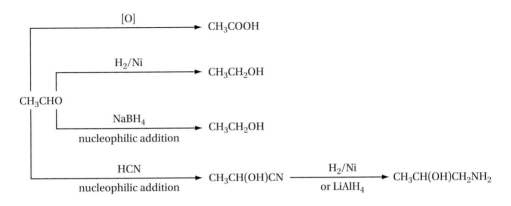

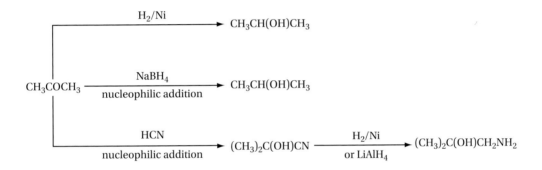

Carboxylic acids and esters

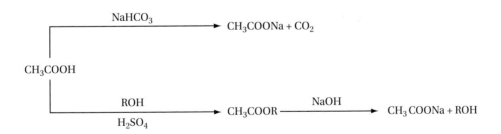

Acylation

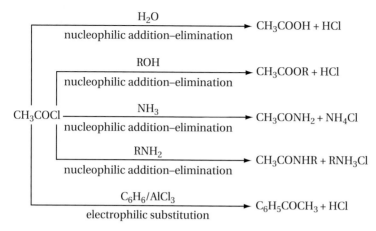

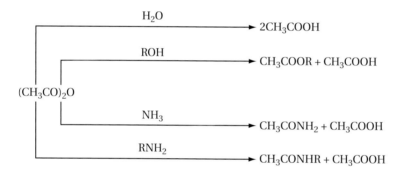

Aromatic chemistry

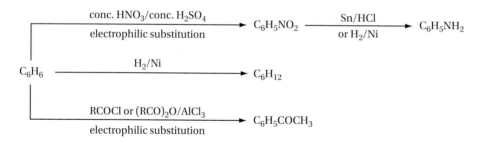

How Science Works

The introduction of the *How Science Works* component into the new A-level specifications has made formal an approach to the teaching of topics in science which many teachers have, in fact, already been using.

Irrespective of future careers, science students need to become proficient in dealing with the various issues included in *How Science Works* so as to achieve a level of scientific awareness. In order to gain an appreciation of how chemists, in particular, work (using the scientific method) it is necessary to understand and be able to apply the concepts, principles and theories of chemistry. The ways that chemical theories and laws are developed, together with the potential impact of new discoveries on society in general, should become clear to you as new concepts in the specification are explored.

Science starts with experimental observation and investigation, followed by verification (i.e. confirmation by others that the results are reliable). A theory or model is proposed to try to explain a set of observations, which may themselves have been accidental or planned as part of a series of experiments. During A-level chemistry courses, many different types of experiment will be carried out and evaluated.

Once an initial theory, or hypothesis, has been put forward to explain a set of results, further experiments are carried out to test the ability of this theory to make accurate predictions. An initial theory may need to be adapted to take into account fresh evidence. More observations and experimental results are produced and the cycle is continued until a firm theory can be established. It is very important that all experiments are carried out objectively, without any bias towards a desired result.

A-level Chemistry courses provide various opportunities for students to analyse verified experimental data as well as the chance to develop theories based on novel findings. In some instances, it will be recognised that there is insufficient experimental evidence for a firm theory to be accepted. In other cases, different experiments carried out by different people will produce contradictory results. Sometimes an apparently unusual result or observation, if verified, will be of great significance. There then follow opportunities to debate these issues and to understand how conflicting theories can be resolved by the accumulation of further evidence.

The concepts and principles dealt with under *How Science Works* will be assessed during the examination. The questions set will require only a knowledge and understanding of the topics included in the specification. In some instances, however, the ability to analyse and make deductions from unfamiliar information may be required. Typical examination questions are provided at the end of the Unit and aspects of *How Science Works* are highlighted.

Practice exam-style questions

1 A slow hydrolysis reaction occurs when methyl propanoate is added to water. However, when methyl propanoate is added to hydrochloric acid, the ester undergoes more rapid hydrolysis. In the table below, data are given for the initial rate for this hydrolysis measured at a fixed temperature with different starting concentrations of acid.

Initial concentration of methyl propanoate/mol dm^{-3}	0.68	0.68
Initial concentration of hydrochloric acid/mol dm^{-3}	0.50	0.73
Initial rate/mol dm^{-3} s^{-1}	4.62×10^{-4}	6.75×10^{-4}

A rate equation for the hydrolysis was determined in further experiments and was found to be:

$$rate = k[\text{methyl propanoate}][\text{hydrochloric acid}]$$

(a) Write an equation for the hydrolysis of methyl propanoate.

_____

_____ 1 mark

(b) Show that the reaction rates given above support the conclusion that the reaction is first order in hydrochloric acid.

_____

_____

_____

_____ 2 marks

(c) Calculate the value of the rate constant, k, for this hydrolysis and state its units.

_____

_____

_____

_____ 2 marks

(d) What evidence do these experiments provide to support the conclusion that hydrochloric acid

(i) is acting as a catalyst?

_____

_____

(ii) is acting as a homogeneous catalyst?

_____

_____ 3 marks

(e) The mechanism of acid-catalysed ester hydrolysis has been studied and it has been suggested that the first step could involve the initial addition of a proton to the ester to form a carbocation, as in *Step 1* below.

Step 1 $RCOOR' + H^+ \rightleftharpoons RC^+(OH)OR'$

(i) Write an equation for the reaction of this carbocation intermediate with water. This is *Step 2*.

_____

_____

(ii) Write the overall equation that results from combining *Step 1* and *Step 2*.

_____

_____

(iii) Which one of these two steps is the rate-determining step? Give a reason for choosing this step and also a reason for rejecting the other step.

Rate-determining step _____

Reason for choosing this step _____

_____

Reason for rejecting the other step _____

_____ 5 marks

Total Marks: 13

2 In the Contact Process for the production of sulfuric acid, sulfur dioxide and oxygen react to form sulfur trioxide in the presence of vanadium(V) oxide. Temperatures in excess of 500 K and pressures in the region of 1 MPa are commonly used. The equation for the reaction is:

$$2SO_2(g) + O_2(g) \rightleftharpoons 2SO_3(g) \qquad \Delta H^{\ominus} = -190 \text{ kJ mol}^{-1}$$

(a) (i) What would be the effect on the yield of SO_3 of an increase in the temperature? Explain your answer.

Effect on yield of increased temperature _____

Explanation _____

(ii) What would be the effect of an increase in temperature on the value of the equilibrium constant for this process?

_____ 4 marks

(b) What would be the effect on the yield of SO_3 of an increase in the pressure? Explain your answer.

Effect on yield of increased pressure ———————————————————————————

Explanation ————————————————————————————————

————————————————————————————————————— 3 marks

(c) Manufacturers of sulfuric acid by the Contact Process use high temperatures and moderate pressures. In the light of your answers to parts (a) and (b) above, suggest why conditions that maximise the yield are not chosen.

Reasons for choice of temperature ——————————————————————————

——

——

Reasons for choice of pressure ————————————————————————————

——

————————————————————————————————— 4 marks

(d) Suggest a method, other than altering the equilibrium temperature and pressure, that could be used to increase the yield of SO_3 at equilibrium.

————————————————————————————————— 1 mark

(e) Write an expression for the equilibrium constant, K_c, for this reaction.

——

——

————————————————————————————————— 1 mark

(f) In a test experiment carried out in a vessel of volume 0.48 dm^3, it was found that equal numbers of moles of SO_2 and SO_3 were present in an equilibrium mixture that also contained 0.75 moles of oxygen.

Calculate the value of the equilibrium constant, K_c, for this reaction and state its units.

——

——

——

——

——

————————————————————————————————— 4 marks

Total Marks: 17

3 **(a)** Titrations were carried out using the following solutions:

Solution **B**, containing 0.80 mol dm^{-3} of sodium hydroxide

Solution **S**, containing 1.0 mol dm^{-3} of hydrochloric acid

Solution **W**, containing 1.0 mol dm^{-3} of ethanoic acid

[Assume that $K_w = 1.0 \times 10^{-14}$ mol^2 dm^{-6}]

(i) Calculate the volume of **B** that is required to neutralise a 20.0 cm^3 portion of **S**.

_____

_____

(ii) State the pH of the neutralised solution at the equivalence point.

_____

(iii) Determine the pH of the solution that results when 50.0 cm^3 of **B** are added to a 20.0 cm^3 portion of **S**.

_____

_____

_____

_____

_____

_____ 7 marks

(b) Titration curves are drawn when separate 20 cm^3 portions of **S** and **W** are titrated with **B**.

(i) State two features (such as volume, pH value or shape) that are exactly the same in the two titration curves.

Identical feature 1 _____

Identical feature 2 _____

(ii) State two features (such as volume, pH value or shape) that are distinctly different in the titration curves.

Different feature 1 _____

Different feature 2 _____ 4 marks

(c) Solution **P** contains n moles of a weak acid, HA. The addition of some sodium hydroxide to **P** neutralises one quarter of the HA present to produce solution **Q**, in which $[H^+] = 7.2 \times 10^{-4}$ mol dm^{-3}.

(i) In terms of the number of moles, n, how many moles of HA remain in **Q**?

_____

_____

(ii) Determine the ratio $\dfrac{[HA]}{[A^-]}$ in **Q**.

_____

_____

(iii) Use your answer to part (ii) above to determine the value of the acid dissociation constant of HA.

_____

_____

_____

_____ 5 marks

Total Marks: 16

4 **(a)** List the characteristic features of a *buffer* solution.

_____

_____

_____ 3 marks

(b) (i) State what is meant by the term *acidic buffer* and give the reagents you would choose to make such a buffer. Write an equation for the equilibrium established in a buffer solution containing these reagents.

Acidic buffer _____

Reagents _____

_____

Equation _____

_____

(ii) Use this equation to illustrate how the acidic buffer works.

_____

_____

_____

_____

(iii) State what is meant by the term *basic buffer* and give the reagents you would choose to make such a buffer. Write an equation for the equilibrium established in a buffer solution containing these reagents.

Basic buffer _____

Reagents _____

_____

Equation _____

_____ 10 marks

(c) What is the pH of the buffer solution that results from the addition of 12.0 cm^3 of 0.20 mol dm^{-3} NaOH to 20.0 cm^3 of 0.24 mol dm^{-3} of a weak acid HX (pK_a = 5.00)?

_____

_____

_____

_____

_____ 6 marks

Total Marks: 19

5 **(a)** Name the type of stereoisomerism shown by 2-hydroxypropanoic (*lactic*) acid, CH$_3$CH(OH)COOH. Identify the structural feature of the molecule that permits the existence of two isomers. With the aid of diagrams, illustrate the structural relationship between these isomers. Explain how it is possible to distinguish between the two isomers.

Type of stereoisomerism _____

Structural feature _____

_____

          Isomer 1   *Isomer 2*

Explanation _____

_____ 6 marks

(b) Explain, with reference to the isomers of lactic acid, the meaning of the terms *enantiomer* and *racemate*.

Enantiomer _____

_____

_____

Racemate _____

_____

_____ 4 marks

(c) Explain why the melting point of (+)-2-aminopropanoic acid (*alanine*), $CH_3CH(NH_2)COOH$ (314 °C), is much higher than that of (+)-2-hydroxypropanoic acid (53 °C).

_____

_____

_____

_____ 4 marks

(d) When an aqueous solution of alanine is electrolysed at low pH, an organic species is attracted to the cathode. At high pH, a different organic species is attracted to the anode. Deduce the structure of each of the moving species.

Species moving to cathode at low pH　　　　　　　*Species moving to anode at high pH*

2 marks

(e) Using RNH_2 to represent alanine, write an equation for the reaction between alanine and ethanoyl chloride. Name and outline the mechanism of the reaction.

Equation _____

_____

Name of mechanism _____

Mechanism

6 marks

Total Marks: 22

6 **(a)** Ketone **K** has the molecular formula C_4H_8O.

(i) The mass spectrum of compound **K** contains a major peak at $m/z = 43$. Deduce the structure of the fragment ion responsible for this peak and write an equation showing its formation from the molecular ion of **K**.

Fragment ion at m/z = 43 _____

Equation _____

(ii) Less intense, but significant, peaks appear at $m/z = 57$ and $m/z = 29$, respectively. Deduce the structures of the fragment ions that are responsible for these two peaks.

Fragment ion at m/z = 57 _____

Fragment ion at m/z = 29 _____

(iii) Give the structural formula and name of ketone **K**.

Structural formula of **K** _____

Name _____ 6 marks

(b) Aldehyde **A** also has the molecular formula C_4H_8O.

Unlike the ^{13}C n.m.r. spectrum of ketone **K**, which has four peaks, the ^{13}C n.m.r. spectrum of aldehyde **A** has only three peaks. Explain why aldehyde **A** does not have four peaks. Give the structural formula and name of aldehyde **A**.

Explanation _____

_____

Structural formula of **A** _____

Name _____ 3 marks

(c) Aldehyde **A** and ketone **K** can be distinguished by using a simple chemical test.

Name the reagent you would use for this test and describe what you would observe. Give the structural formula of the organic reaction product.

Name of reagent _____

Observation _____

Structural formula of reaction product _____

_____ 3 marks

(d) Using RCHO to represent aldehyde **A**, write an equation for the reaction between the aldehyde and hydrogen cyanide. Name and outline the mechanism of the reaction. Explain briefly why special care has to be taken when carrying out such a reaction.

Equation _____

Name of mechanism _____

Mechanism

Explanation _____

_____ 5 marks

(e) Explain why the product obtained in the reaction in part (d) is formed as an optically-inactive racemate.

_____

_____ 2 marks

Total Marks: 19

7 Tallow, obtained from beef fat, can be used to make soap and also biodiesel. A representative structure of tallow is shown below.

$$CH_2OOC(CH_2)_{14}CH_3$$
$$|$$
$$CHOOC(CH_2)_{16}CH_3$$
$$|$$
$$CH_2OOC(CH_2)_7CH{=}CH(CH_2)_7CH_3$$

(a) What type of process is used to convert a fat of this type into a soap? Give the reagent that would be used for this purpose. Name the neutral organic by-product of this reaction.

Type of process _____

Reagent _____

Name of organic by-product _____

_____ 3 marks

(b) (i) Unsaturated groups present in oils and fats have the *Z*-configuration in most cases. Using RCH=CHR, draw a structure to show what is meant by a *Z*-configuration. Name the alternative type of configuration.

Structure

Alternative type _____

(ii) Give a reagent system that can be used to convert an unsaturated group into a saturated group.

_____ 4 marks

(c) The formation of biodiesel from oils and fats involves heating them with an excess of a simple alcohol. Name this type of process and give the structure of one of the esters present in biodiesel that is obtained by heating tallow with an excess of methanol. Explain briefly why it is necessary for an excess of the alcohol to be present.

Type of process _____

Structure of ester _____

Explanation _____

_____

_____ 4 marks

(d) Terylene is made by heating together ethane-1,2-diol and dimethyl benzene-1,4-dicarboxylate. Gaseous methanol is formed as a by-product. Draw the repeating unit of Terylene and explain why this polymer, like biodiesel, is biodegradable.

Repeating unit

Explanation _____

_____ 4 marks

Total Marks: 15

8 The local anaesthetic benzocaine (**S**) can be made by the following sequence.

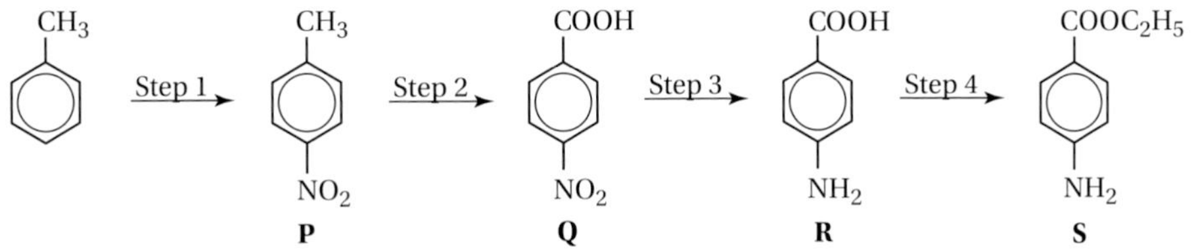

(a) In *Step 1*, methylbenzene is treated with a mixture of two reagents. Identify the two reagents and write an equation showing the formation of the reactive species involved. State the type of reaction taking place, name and outline the mechanism.

Reagent 1 ——————————————— *Reagent 2* ———————————————————

Equation ——————————————————————————————————————

Type of reaction ———————————————————————————————————

Name of mechanism ————————————————————————————————

Mechanism

7 marks

(b) The product obtained from *Step 1* is a liquid and contains compound **P** contaminated with other isomers. When pure, compound **P** is a crystalline solid, m.p. 51 °C. Describe briefly how chromatography could be used to obtain a pure sample of crystalline **P** from 5 g of the impure liquid mixture.

———

———

———

———

——— 5 marks

(c) Name the type of reaction occurring in *Step 2*.

——— 1 mark

(d) Name the type of reaction taking place in *Step 3* and give a suitable reagent or mixture of reagents for this conversion.

Type of reaction ———————————————————————————————————

Reagent(s) ——————————————————————————————— 3 marks

(e) Compound **R** contains a small amount of unreacted compound **Q**. Both compound **R** and compound **Q** are solids which are sparingly soluble in water. Suggest a way of obtaining compound **R** as a pure sample and explain the basis of your method.

Method of separation _____

_____

Explanation _____ 4 marks

(f) Name the type of reaction occurring in *Step 4* and identify the reagents needed for this conversion. Give the systematic name of compound **S**.

Type of reaction _____

Reagents _____

Name _____

_____ 4 marks

Total Marks: 24

9 **(a)** Write an equation to illustrate how the primary amine RNH_2 functions as a Brønsted–Lowry base. Explain why ethylamine (pK_a 10.73), $C_2H_5NH_2$, is a stronger base than ammonia (pK_a 9.25), whereas phenylamine (pK_a 4.62), $C_6H_5NH_2$, is less basic than ammonia.

Equation _____

Explanation _____

_____

_____

_____ 5 marks

(b) Amines such as propylamine, $CH_3CH_2CH_2NH_2$, can be made from haloalkanes and also from nitriles. Name the type of reaction involved and the reagent(s) used for the conversion of (i) 1-bromopropane and (ii) propanenitrile into propylamine. In each case, write an equation for the reaction.

(i) *Type of reaction* _____

Reagent(s) _____

Equation _____

(ii) *Type of reaction* _____

Reagent(s) _____

Equation _____ 6 marks

(c) Name and outline a mechanism for reaction (b)(i) and explain why this reaction is a less satisfactory method for the preparation of propylamine than reaction (b)(ii).

Name of mechanism _____

Mechanism

Explanation _____

_____

_____ 5 marks

(d) Give the structure and name the type of compound formed when R_2NH is heated with a large excess of chloromethane. Give one use of the product obtained when R is a very long alkyl group.

Structure

Type of compound _____

Use _____ 4 marks

Total Marks: 20

10 The —CONH— unit occurs in various types of organic compounds, ranging from simple amides and cyclic structures to synthetic polymers and complex proteins.

(a) Caprolactam, the cyclic amide of 6-aminohexanoic acid, is the monomer used in a ring-opening polymerisation reaction to manufacture nylon-6.

(i) Draw the structure of caprolactam and give the repeating unit of nylon-6.

Structure of caprolactam

Repeating unit of nylon-6 _____

(ii) Deduce the structure of the product obtained when caprolactam is hydrolysed with aqueous sodium hydroxide.

3 marks

(b) The aromatic polyamide Kevlar is obtained from the reaction between benzene-1,4-dicarboxylic acid and benzene-1,4-diamine. Draw the repeating unit of Kevlar and explain why this polymer has a sheet-like structure.

Repeating unit of Kevlar

Explanation _____

_____ 3 marks

(c) The hydrolysis of a protein molecule, such as insulin, yields a mixture of amino acids. Suggest a laboratory technique suitable for the separation of the components present in a mixture of amino acids. Suggest a means of identification of the various components present in such a mixture.

Laboratory technique _____

Means of identification _____

_____ 2 marks

(d) A dipeptide is formed as the simplest combination of two amino acids. Deduce the number of different dipeptides produced when glycine, H_2NCH_2COOH, and alanine, $H_2NCH(CH_3)COOH$, are heated together. Give the structures of two of the possible isomers.

Number of dipeptides formed _____

Isomer 1 _____

_____

Isomer 2 _____

_____ 3 marks

Total Marks: 11

11 Various polymers have monomers which can be represented by $XCH{=}CH_2$. Examples include poly(propene) (X = CH_3) and poly(chloroethene) (X = Cl). Typical uses of these products include tubing and food containers. Used products could end up on landfill sites, be recycled or incinerated.

(a) State the type of polymer formed by the monomers given above and draw the repeating unit of poly(chloroethene).

Type of polymer _____

Repeating unit _____

_____ 2 marks

(b) Explain briefly why neither of these polymers is biodegradable.

_____

_____ 2 marks

(c) Suggest one advantage and one disadvantage of recycling poly(propene).

Advantage _____

Disadvantage _____ 2 marks

(d) Suggest one advantage and one disadvantage of incinerating poly(chloroethene).

Advantage _____

Disadvantage _____ 2 marks

Total Marks: 8

12 A mixture of two organic liquids could not be separated efficiently using fractional distillation. Use of a different laboratory technique permitted the isolation of compound **X**, b.p. 202 °C, and compound **Y**, b.p. 205 °C.

The mass spectrum of compound **X** has the molecular ion peak at $m/z = 120$ together with two major peaks at $m/z = 77$ and $m/z = 105$, respectively. A dominant peak appears in the infra-red spectrum of **X** at 1690 cm^{-1}. One peak in the ^1H n.m.r. spectrum of **X** is a three-proton singlet at δ 2.60 ppm and one of the peaks present in the ^{13}C n.m.r. spectrum has a strong signal at δ 198 ppm. Compound **X** remains unchanged when heated with acidified potassium dichromate(VI) solution.

The mass spectrum of compound **Y** has the molecular ion peak at $m/z = 108$ and a major peak at $m/z = 77$. An intense peak appears in the infra-red spectrum of **Y** at 3352 cm^{-1}. One peak in the ^1H n.m.r. spectrum of **Y** is a two-proton singlet at δ 4.50 ppm. Compound **Y** is oxidised to benzenecarboxylic acid when heated with acidified potassium dichromate(VI) solution.

(a) Suggest which technique was used to separate compounds **X** and **Y**.

_____ 1 mark

(b) What conclusions can be drawn from the fact that both **X** and **Y** have a fragment ion at $m/z = 77$ in their mass spectra?

_____

_____ 2 marks

(c) By reference to the various m/z values provided, deduce the structure of the fragment ion appearing at $m/z = 105$ in the mass spectrum of compound **X**.

_____

_____ 2 marks

(d) For compound **X**, identify the groups associated with the following peaks.

1690 cm^{-1} in the infra-red spectrum

_____

δ 2.60 ppm in the ^1H n.m.r. spectrum

_____

δ 198 ppm in the ^{13}C n.m.r. spectrum

_____ 3 marks

(e) Deduce the structure of compound **X**.

_____ 1 mark

_____

(f) For compound **Y**, identify the groups associated with the following peaks.

3352 cm^{-1} in the infra-red spectrum

_____

δ 4.50 ppm in the ^{1}H n.m.r. spectrum

_____ 2 marks

(g) Deduce the structure of compound **Y**.

_____ 1 mark

(h) Explain how the behaviour of compounds **X** and **Y** towards acidified potassium dichromate(VI) solution relates to the structures of these compounds.

_____

_____ 2 marks

Total Marks: 14

^{1}H n.m.r. chemical shift data		^{13}C n.m.r. chemical shift data		Infra-red absorption data	
Type of proton	δ /ppm	Type of carbon	δ /ppm	Bond	Wavenumber/cm^{-1}
RCH$_3$	0.7–1.2	RCH$_3$	5–30	C – H	2850–3300
R$_2$CH$_2$	1.2–1.4	R$_2$CH$_2$	15–40	C – C	750–1100
R$_3$CH	1.4–1.6	R$_3$CH	20–40	C = C	1620–1680
RCOCH$_3$	2.1–2.6	CC=O	20–50	C = O	1680–1750
ROCH$_3$	3.1–3.9	COH	50–75	C – O	1000–1300
RCOOCH$_3$	3.7–4.1	C=O (ester, acid)	160–185	O – H (alcohols)	3230–3550
ROH	0.5–5.0	C=O (aldehyde, ketone)	190–220	O – H (acids)	2500–3000

Answers, explanations, hints and tips

Question	Answer		Marks
1 (a)	$CH_3CH_2COOCH_3 + H_2O \rightleftharpoons CH_3CH_2COOH + CH_3OH$	(1)	1
1 (b)	ratio of HCl concentrations = 0.73/0.50 = 1.46	(1)	
	ratio of rates = 6.75/4.62 = 1.46	(1)	2
1 (c)	$k = \dfrac{4.62 \times 10^{-4}}{0.68 \times 0.50} = 1.36 \times 10^{-3}$	(1)	
	$dm^3\ mol^{-1}\ s^{-1}$	(1)	2
1 (d) (i)	speeds up reaction	(1)	
	in rate equation but not in stoichiometric equation (or not consumed)	(1)	
1 (d) (ii)	acts in same phase as reactants	(1)	3
1 (e) (i)	$RC^+(OH)OR' + H_2O \rightarrow RCOOH + R'OH + H^+$	(1)	
1 (e) (ii)	$RCOOR' + H_2O \rightleftharpoons RCOOH + R'OH$	(1)	
1 (e) (iii)	step 1	(1)	
	step 1 involves both substances that appear in the rate equation	(1)	
	step 2 involves water, which does not appear in the rate equation	(1)	5
			Total 13
2 (a) (i)	decreases	(1)	
	exothermic reaction	(1)	
	equilibrium shifts to left to decrease temperature rise	(1)	
2 (a) (ii)	decreases	(1)	4
2 (b)	increases	(1)	
	number of molecules decreases left to right	(1)	
	equilibrium shifts to right to decrease pressure rise	(1)	3
2 (c)	low temperatures mean low rates of reaction	(1)	
	compromise chosen between rate and yield	(1)	
	high pressures expensive to produce	(1)	
	moderate pressure chosen	(1)	4
2 (d)	removal of product as it is formed *or* recycle gases after product removal	(1)	1
2 (e)	$K_c = \dfrac{[SO_3]^2}{[SO_2]^2\,[O_2]}$	(1)	1
2 (f)	$\dfrac{[SO_3]^2}{[SO_2]^2} = 1$ when both present in equimolar amounts	(1)	
	so $K_c = \dfrac{1}{[O_2]}$	(1)	
	$= \dfrac{1}{0.75/0.48} = 0.64$	(1)	
	$dm^3\ mol^{-1}$	(1)	4
			Total 17
3 (a) (i)	$V = \dfrac{20.0 \times 1.0}{0.80} = 25.0\ cm^3$	(1)	
3 (a) (ii)	pH = 7	(1)	

Question	Answer		Marks
3 (a) (iii)	$50.0 - 25.0 = 25.0$ cm^3 of 0.80 mol dm^{-3} NaOH in excess	(1)	
	$= \dfrac{25.0 \times 0.80}{1000}$ = 0.020 mol OH$^-$ excess in a volume of 70.0 cm^3	(1)	
	so [OH$^-$] $= \dfrac{0.020 \times 1000}{70.0}$ = 0.286 mol dm^{-3}	(1)	
	and [H$^+$] $= \dfrac{K_w}{[\text{OH}^-]}$ $= \dfrac{1.0 \times 10^{-14}}{0.286}$ = 3.50×10^{-14} mol dm^{-3}	(1)	
	so pH = 13.46	(1)	7
3 (b) (i)	equivalence point (25 cm^3) same for both	(1)	
	curve shape and final pH identical after equivalence point	(1)	
3 (b) (ii)	ethanoic acid curve starts at higher pH and lies above HCl curve	(1)	
	equivalence pH > 7 for ethanoic acid	(1)	4
3 (c) (i)	$\dfrac{3}{4}n$	(1)	
3 (c) (ii)	$\left(\dfrac{3}{4}n\right)\Big/\left(\dfrac{1}{4}n\right) = 3$	(1)	
3 (c) (iii)	$K_a = \dfrac{[\text{H}^+][\text{A}^-]}{[\text{HA}]}$	(1)	
	$= \dfrac{[\text{H}^+]}{3}$	(1)	
	= 2.4 × 10^{-4} mol dm^{-3}	(1)	5
			Total 16
4 (a)	resists change in pH on addition of small amounts of acid	(1)	
	or of base	(1)	
	and on dilution	(1)	3
4 (b) (i)	one with a pH less than 7	(1)	
	a solution of a weak acid and one of its salts *or* a part-neutralised weak acid	(1)	
	e.g. CH$_3$COOH(aq) $\rightleftharpoons$ H$^+$(aq) + CH$_3$COO$^-$(aq)	(1)	
4 (b) (ii)	on adding H$^+$, excess reacts with CH$_3$COO$^-$ to form CH$_3$COOH	(1)	
	on adding OH$^-$, excess reacts with CH$_3$COOH *or* with H$^+$ to form more CH$_3$COO$^-$	(1)	
	both of these reactions decrease the effect of the change in pH	(1)	
	on dilution, proportions of CH$_3$COOH and CH$_3$COO$^-$ do not alter, so [H$^+$] does not change	(1)	
4 (b) (iii)	one with a pH greater than 7	(1)	
	a solution of a weak base and one of its salts *or* a part-neutralised weak base	(1)	
	e.g. NH$_3$(aq) + H$_2$O(l) $\rightleftharpoons$ NH$_4^+$(aq) + OH$^-$(aq)	(1)	10
4 (c)	$K_a = \dfrac{[\text{H}^+][\text{X}^-]}{[\text{HX}]}$ so $\qquad [\text{H}^+] = K_a \dfrac{[\text{HX}]}{[\text{X}^-]}$	(1)	
	original moles HX $= \dfrac{20.0}{1000} \times 0.24$ $\qquad = 0.0048$ mol	(1)	
	OH$^-$ added $= \dfrac{12.0}{1000} \times 0.20$ $\qquad = 0.0024$ mol		
	so moles X$^-$ formed $\qquad\qquad = 0.0024$ mol	(1)	
	HX remaining $\quad = 0.0048 - 0.0024 \quad = 0.0024$ mol	(1)	
	pH $= pK_a - \log\dfrac{[\text{HX}]}{[\text{X}^-]}$ $\qquad = 5.00 - \log 1$	(1)	
	$= 5.00$	(1)	6
			Total 19

Question	Answer		Marks
5 (a)	optical	(1)	
	an (asymmetric) carbon atom with four different groups attached	(1)	
	[a chiral molecular structure]		
		(1)	
	[the structures should show a mirror-image relationship]		
	(equal but) opposite rotation	(1)	
	of plane-polarised light	(1)	6
5 (b)	Isomer 1 and Isomer 2 are enantiomers	(1)	
	each is a non-superimposable mirror image of the other	(1)	
	a mixture of equal amounts of Isomer 1 and Isomer 2 is a racemate	(1)	
	optically inactive [50% (+) and 50% (–)]	(1)	4
5 (c)	only hydrogen bonding is possible in (+)-lactic acid	(1)	
	(+)-alanine exists as a zwitterion (dipolar ion)	(1)	
	$H_3N^+CHCOO^-$ CH_3	(1)	
	strong ionic bonding	(1)	4
5 (d)	$H_3N^+CHCOOH$ to cathode CH_3	(1)	
	$H_2NCHCOO^-$ to anode CH_3	(1)	2
5 (e)	$2RNH_2 + CH_3COCl \rightarrow RNHCOCH_3 + RNH_3Cl$	(1)	
	(nucleophilic) addition–elimination	(1)	
	arrows and lone pair (1) structure (1)		6
		Total 22	
6 (a) (i)	$CH_3\overset{+}{C}{=}O$	(1)	
	$C_4H_8O^{+\bullet} \rightarrow CH_3CO^+ + C_2H_5{}^\bullet$	(1)	
6 (a) (ii)	$CH_3CH_2CO^+$	(1)	
	$CH_3CH_2{}^+$	(1)	
6 (a) (iii)	$CH_3COCH_2CH_3$	(1)	
	butan-2-one or butanone	(1)	6

Question	Answer		Marks
6 (b)	two of the carbon atoms are equivalent	(1)	
	$(CH_3)_2CHCHO$	(1)	
	2-methylpropanal *or* methylpropanal	(1)	3
6 (c)	Fehling's solution *or* Tollens' reagent	(1)	
	A gives a red precipitate from a deep blue solution *or* a silver mirror	(1)	
	$(CH_3)_2CHCOOH$	(1)	3
6 (d)	$RCHO + HCN \rightarrow RCH(OH)CN$	(1)	
	nucleophilic addition	(1)	
	HCN is highly toxic (and causes death by inhalation)	(1)	5
6 (e)	carbonyl group is planar	(1)	
	attack by cyanide ion equally likely at either side	(1)	2
			Total 19
7 (a)	hydrolysis *or* saponification	(1)	
	aqueous sodium hydroxide	(1)	
	propane-1,2,3-triol *or* glycerol	(1)	3
7 (b) (i)	both R groups on same side of double bond	(1)	
	E-configuration	(1)	
7 (b) (ii)	H_2	(1)	
	Ni catalyst	(1)	4
7 (c)	transesterification	(1)	
	$CH_3(CH_2)_{14}COOCH_3$ *or* $CH_3(CH_2)_{16}COOCH_3$ *or*		
	$CH_3(CH_2)_7CH{=}CH(CH_2)_7COOCH_3$	(1)	
	drive equilibrium to right	(1)	
	to oppose change imposed (Le Chatelier's principle)	(1)	4
7 (d)		(2)	
	repeating unit can be split by hydrolysis	(1)	
	by enzyme action	(1)	4
			Total 15
8 (a)	conc. H_2SO_4 + conc. HNO_3	(1)	
	$HNO_3 + 2H_2SO_4 \rightarrow NO_2^+ + H_3O^+ + 2HSO_4^-$	(1)	
	nitration	(1)	
	electrophilic substitution	(1)	
	structure (1)		7

Question	Answer		Marks
8 (b)	dissolve in a solvent, e.g. methylbenzene	(1)	
	use column chromatography	(1)	
	on alumina *or* silica	(1)	
	collect main fraction	(1)	
	remove solvent and recrystallise	(1)	5
8 (c)	oxidation	(1)	1
8 (d)	reduction *or* hydrogenation	(1)	
	H_2	(1)	
	Ni catalyst	(1)	3
8 (e)	dissolve in aqueous HCl	(1)	
	filter off compound **Q**	(1)	
	add aqueous NaOH to precipitate **R**	(1)	
	basic amino group in **R**	(1)	4
8 (f)	esterification	(1)	
	ethanol	(1)	
	acid catalyst	(1)	
	ethyl 4-aminobenzenecarboxylate	(1)	4
			Total 24
9 (a)	$R\ddot{N}H_2 + H^+ \rightleftharpoons RNH_3^+$	(1)	
	inductive effect of ethyl group pushes electrons towards N	(1)	
	increases electron density on N (lone pair more available)	(1)	
	delocalisation of N lone pair in phenylamine	(1)	
	reduces electron density on N (lone pair less available)	(1)	5
9 (b) (i)	alkylation	(1)	
	NH_3	(1)	
	$CH_3CH_2CH_2Br + 2NH_3 \rightarrow CH_3CH_2CH_2NH_2 + NH_4^+Br^-$	(1)	
9 (b) (ii)	hydrogenation *or* reduction	(1)	
	Ni/H_2 *or* $LiAlH_4$	(1)	
	$CH_3CH_2CN + 2H_2 \rightarrow CH_3CH_2CH_2NH_2$	(1)	6
9 (c)	nucleophilic substitution	(1)	
	mixture obtained due to further substitution	(1)	
	only one product by reduction	(1)	5
9 (d)	$[R_2N(CH_3)_2]^+Cl^-$		
	(1) (1)		
	quaternary ammonium salt	(1)	
	cationic surfactant	(1)	4
			Total 20

Question	Answer		Marks
10 (a) (i)		(1)	
10 (a) (ii)	$-CO(CH_2)_5NH-$ $H_2NCH_2CH_2CH_2CH_2CH_2COONa$	(1) (1)	3
10 (b)	 hydrogen bonding between chains forms a two-dimensional structure	(1) (1) (1)	3
10 (c)	paper *or* thin-layer chromatography comparison with known standards	(1) (1)	2
10 (d)	4 $H_2NCH_2CONHCH_2COOH$ *or* $H_2NCH(CH_3)CONHCH(CH_3)COOH$ $H_2NCH_2CONHCH(CH_3)COOH$ *or* $H_2NCH(CH_3)CONHCH_2COOH$	(1) (1) (1)	3
			Total 11
11 (a)	addition *or* chain-growth polymers $-CH_2-CH-$ $\quad\quad\ \ \|$ $\quad\quad\ \ Cl$	(1) (1)	2
11 (b)	chemically inert *or* devoid of functional groups cannot be split by biological organisms (enzymes)	(1) (1)	2
11 (c)	conserves natural resources (crude oil) recycled polymer may not be suitable for original purpose	(1) (1)	2
11 (d)	source of energy *or* reduces bulk liberates HCl	(1) (1)	2
			Total 8
12 (a)	gas–liquid *or* column chromatography	(1)	1
12 (b)	$m/z = 77$ corresponds to $C_6H_5^+$ both **X** and **Y** are monosubstituted benzene derivatives	(1) (1)	2
12 (c)	$105 = 120 - 15\ (CH_3) = 77 + 28\ (C=O)$ $C_6H_5CO^+$	(1) (1)	2
12 (d)	C=O group uncoupled CH_3 group C=O group	(1) (1) (1)	3
12 (e)	$C_6H_5COCH_3$	(1)	1
12 (f)	alcohol OH group uncoupled CH_2 group	(1) (1)	2

Question	Answer		Marks
12 (g)	$C_6H_5CH_2OH$	(1)	1
12 (h)	$COCH_3$ not easily oxidised	(1)	
	CH_2OH oxidised to COOH via CHO	(1)	2
			Total 14

The table below highlights aspects of *How Science Works* in the exemplar questions.

Question	How Science Works
1 (a)/(b)/(c)	analysis and interpretation of experimental data
1 (d)/(e)	use of theory to explain observations
2 (c)	data evaluation to make a choice of compromise conditions
2 (f)	analysis and interpretation of experimental data
3	analysis and interpretation of experimental data
5 (d)	interpretation of experimental evidence
6 (a)/(b)	analysis of spectral data
7 (d)	biodiesel; biodegradability
8	explanation of practical activities and prediction of method needed
9 (c)	comparison of synthetic methods
10 (c)	choice of analytical methods
11	advantages and disadvantages of biodegradable polymers and recycling
12	analysis and interpretation of experimental and spectroscopic evidence

Glossary

acid–base equilibrium	the equilibrium transfer of a proton from an acid to a base
acid dissociation constant (K_a)	the equilibrium constant for the dissociation of a weak acid in water; for the weak acid HA, $K_a = \dfrac{[H^+(aq)][A^-(aq)]}{[HA(aq)]}$
acidic buffer	one that maintains pH at a value below 7
activation energy	the minimum energy required for a reaction to occur
acyl group	a functional group (RC=O) derived from a carboxylic acid
acylation	the introduction of an acyl group into an organic molecule
acylium cation	the electrophile $[RCO]^+$
addition polymer	one obtained by the addition of monomers to the end of a growing chain
adsorption chromatography	involves a solid phase of finely-divided particles as the fixed (stationary) phase and a liquid or a gas as the moving (mobile) phase
amino acid	the name commonly used for compounds having a primary amino group attached to the carbon atom adjacent to a carboxylic acid group
analytical chromatography	operates with small amounts of material and aims to identify and measure the relative proportions of the various components present in a mixture
arenes	monocyclic or polycyclic aromatic hydrocarbons, such as benzene or naphthalene
aromatic	the name traditionally used in relation to benzene and its derivatives
asymmetric carbon atom	one with four different atoms or groups attached that is devoid of symmetry
base peak	the largest peak in a mass spectrum
basic buffer	one that maintains pH at a value above 7
bimolecular step	a second-order step in a reaction mechanism
biodegradable	capable of being broken down by micro-organisms (enzymes)
biodiesel	a renewable, non-petroleum-based fuel obtained by transesterification from vegetable oils, such as soya bean and rapeseed oil
Brønsted–Lowry acid	a proton donor
Brønsted–Lowry base	a proton acceptor
buffer range	the pH range over which a weak acid/base can show buffer action
buffer region	the concentration range over which a weak acid/base can show buffer action
buffer solution	one that resists changes in pH on addition of small amounts of acid or base, or on dilution
carrier gas	an *eluent* gas, such as helium, used as the moving phase in gas–liquid chromatography
catalyst	a substance that alters the rate of a reaction without itself being consumed
chain-growth polymer	see *addition polymer*

chain isomers	structural isomers which occur when there is more than one way of arranging the carbon skeleton of a molecule
chemical shift (δ)	in n.m.r., the amount, measured in parts per million (ppm), by which a ^1H or a ^{13}C resonance, for example, is shifted from that of an internal standard
chiral drugs	drugs possessing chiral centres, often single-enantiomer structures
chiral molecule	one that cannot be superimposed on its mirror image
chromatogram	a separated pattern of substances in a mixture, obtained by chromatography
chromatograph	an apparatus used for chromatographic separation of volatile components in a mixture
chromatography	a technique for separating the components of a mixture on the basis of their different affinities for a stationary and for a moving phase
column chromatography	involves a stationery phase of finely-divided alumina or silica gel in a vertical glass tube and an organic solvent as the moving phase
condensation polymer	one involving the loss of small molecules, obtained by the reaction between molecules having two functional groups
delocalisation energy	the increase in stability associated with electron delocalisation
delocalised electrons	electrons that are spread over more than one atom in a molecule, e.g. as in benzene, where six delocalised electrons lie above and below the plane of the hexagonal ring
deshielded	in n.m.r., a nucleus is said to be deshielded when the electron density surrounding it is reduced, giving rise to a downfield shift (larger $δ$ value)
diacidic base	one that forms two moles of hydroxide ions per mole of base, e.g. $Ba(OH)_2$
diazotisation	the conversion of $ArNH_2$ into ArN_2^+
diprotic acid	one that forms two moles of protons per mole of acid, e.g. H_2SO_4
doublet	in n.m.r., a peak that is split into two parts
Ecoflex	a fully biodegradable aliphatic–aromatic co-polyester, used for disposable packaging, based on butane-1,4-diol and benzene-1,4-dicarboxylic acid
electrophile	an electron-seeking species, e.g. a positive ion or the more positive end of a polar molecule, which usually accepts a pair of electrons
electrophilic substitution reaction	mechanistically, an electrophilic addition–elimination reaction resulting in overall substitution, typically involving arenes, e.g. nitration of benzene
eluate	the solution emerging from a chromatographic column
eluent	the solvent used as the moving phase in column chromatography
elution	the process of washing the components of a mixture down a chromatographic column
enantiomers	three-dimensional, non-superimposable molecular structure mirror images
endothermic change	one in which heat energy is taken in
end-point	the point during a titration when the colour of an indicator lies half-way between the acid and base colours, i.e. [HIn] = [In⁻] for the indicator HIn

equilibrium constant (K_c)	the ratio of concentrations of products and reactants raised to the powers of their stoichiometric coefficients; e.g. for the reaction $3A \rightleftharpoons 2B + C$ $\qquad K_c = \dfrac{[B]^2[C]}{[A]^3}$
equivalence point	the point on a titration curve at which stoichiometrically equivalent amounts of acid and base have been mixed together
exothermic change	one in which heat energy is given out
E–Z stereoisomerism	also known as geometrical or *cis–trans* isomerism
E–Z stereoisomers	arise due to restricted rotation about a carbon–carbon double bond when the two pairs of attached substituents can be arranged in two different ways
fibrous proteins	contain long chains of polypeptides which occur in bundles, e.g. keratin
fragmentation	in mass spectrometry, the breakdown of a molecular ion into smaller, positively-charged ions and radicals
free-radical substitution reaction	one in which the hydrogen atom of a C—H bond is replaced by a halogen atom; a chain-reaction mechanism involves attack on a neutral molecule by a radical (halogen atom)
Friedel–Crafts acylation	an electrophilic substitution reaction, involving an acylium cation, resulting in carbon–carbon bond formation
functional group	an atom or group of atoms which, when present in different molecules, results in similar chemical properties
functional group isomers	structural isomers which contain different functional groups
gas–liquid chromatography (GLC)	involves an inert powder coated with a film of a non-volatile liquid, packed in a tube (the stationary phase), and a carrier gas (the moving phase)
glass-transition temperature (T_g)	the temperature at which a polymer changes from a hard and glass-like state to a more flexible and mouldable state
globular proteins	contain long chains of amino acids, soluble in water, which are folded into roughly spherical shapes, e.g. haemoglobin
good leaving group	a stable species which is liberated during an organic chemical reaction
half-equivalence	when exactly one-half of the equivalence volume of a base or acid has been added to an acid or base.
heterogeneous system	one with the species present in different phases
heteronuclear species	molecules composed of more than one type of element, e.g. HCl
homogeneous system	one with all species present in the same phase
homologous series	a family of organic molecules which all contain the same functional group but have an increasing number of carbon atoms; each member can be represented by a general formula, e.g. $C_nH_{2n+1}X$
homonuclear species	diatomic molecules composed of only one type of element, e.g. Cl_2
in vivo	within the human body
incineration	waste-treatment technology involving the combustion of organic materials
indicator	usually a weak organic acid with strongly coloured acid (HIn) and base (In^-) forms

initial rate of reaction	the rate of change of concentration at the start of a reaction
integration trace	in n.m.r., a computer-generated line, superimposed on the spectrum, which measures the relative areas under the various peaks in the spectrum
ionic product of water (K_w)	$K_w = [\text{H}^+(\text{aq})][\text{OH}^-(\text{aq})]$
isoelectric point	the pH at which an amino acid has no net charge
isomers	molecules with the same chemical formula but in which the atoms are arranged differently (see *structural isomerism* and *stereoisomerism*)
Kevlar	a sheet-like polyamide, used in bullet-proof vests, derived from benzene-1,4-dicarboxylic acid and benzene-1,4-diamine
landfill site	an area of land on which rubbish is dumped
Le Chatelier's principle	states that a system at equilibrium will respond to oppose any change imposed upon it
magnetic moment	a measure of the torque exerted on a magnetic system, e.g. a bar magnet, when placed in a magnetic field
mobile phase	see *moving phase*
molecular ion ($\text{M}^{+\bullet}$)	the species formed in a mass spectrometer by the loss of one electron from a molecule
monoacidic base	one that forms one mole of hydroxide ions per mole of base, e.g. NaOH
monoprotic acid	one that forms one mole of protons per mole of acid, e.g. HCl
moving phase	in chromatography, the liquid or gaseous phase that passes through a fixed stationary phase
multiplet	in n.m.r., a peak that is split into many parts
$n + 1$ rule	in n.m.r., signals for protons adjacent to n equivalent neighbours are split into $n + 1$ peaks
nitryl cation	the electrophile $^+\text{NO}_2$
Nomex	the 1,3-linked isomer of Kevlar, used in flame-resistant clothing
nuclear spin	a property that influences the behaviour of certain nuclei, typically ^1H and ^{13}C, in a magnetic field; nuclei possessing even numbers of both protons and neutrons, such as ^{12}C and ^{16}O, lack magnetic properties and do not give rise to n.m.r. signals.
nucleophile	an electron-rich molecule or ion able to donate a pair of electrons
nucleophilic addition reaction	one in which an electron-rich molecule or ion (with a lone pair of electrons) attacks the electron-deficient atom of a polar group, e.g. addition of HCN to an aldehyde or ketone
optical isomers	stereoisomers (enantiomers) which rotate the plane of plane-polarised light equally but in opposite directions
optically active	capable of rotating the plane of plane-polarised light
order of reaction	the sum of the powers of the concentration terms in the rate equation
paper chromatography	involves a thin layer of water adsorbed on to chromatographic paper (the stationary phase) and a solvent or solvent mixture (the moving phase)
parent ion	see *molecular ion*

...tion chromatography	involves a thin, non-volatile liquid film held on the surface of an inert solid or within the fibres of a supporting matrix (the stationary phase) and a liquid or a gas (the moving phase)
peptide link	the –CONH– linking group
pH	logarithmic expression of the proton concentration in aqueous solution $pH = -\log_{10}[H^+(aq)]$
pH at half-equivalence	at half-equivalence, $pH = pK_a$ for a weak acid
pK_a	logarithmic expression of the acid dissociation constant in aqueous solution $pK_a = -\log_{10}K_a$
plasticiser	a substance used to soften plastics and increase flexibility, e.g. dibutyl benzene-1,2-dicarboxylate
position isomers	structural isomers which have the same carbon skeleton and the same functional group(s), but in which the functional groups are joined at different places on the carbon skeleton
preparative chromatography	a form of purification of organic compounds, involving chromatography on a relatively large scale
primary structure	of a protein, is the sequence of amino-acid units present in the polymer
proton acceptor	a substance that accepts protons in a chemical reaction
proton-decoupled spectra	are simplified n.m.r. spectra obtained as the result of removing the interactions between ^{13}C nuclei and any attached protons
proton donor	a substance that donates protons in a chemical reaction
quartet	in n.m.r., a peak that is split into four parts
racemate	a mixture containing equal amounts of both enantiomers
racemic mixture	see *racemate*
radical cation	a positively-charged species which possesses an unpaired electron
rate constant (k)	the constant of proportionality in the rate equation
rate-determining step	the slowest step in a multi-step reaction sequence
rate equation	the relationship between the rate of reaction and concentrations of reactants
rate of reaction	the change in concentration of a substance in unit time
reaction mechanism	a sequence of discrete chemical reaction steps that can be deduced from the experimentally observed rate equation
recycling	the processing of used materials, e.g. glass, paper, textiles, metals and plastics, into new products in order to prevent wastage, to reduce the consumption of raw materials and to lower energy costs
repeating unit	the group of atoms that repeats throughout the length of a polymer chain
resolution	the separation of enantiomers *or* the separation of mixtures of chemicals using chromatography
resonance	a concept used when a single molecule can be approximated by more than one classical Lewis structure, involving single and multiple covalent bonds *or* in n.m.r., the excitation of atomic nuclei in a magnetic field by exposure to electromagnetic radiation of a specific frequency

resonance energy	the increase in stability associated with resonance between Lewis structures
resonance hybrid	a representation of an actual molecule, e.g. benzene, when classical structures using single and multiple covalent bonds are inadequate
retention factor (R_f)	in chromatography, $R_f = \dfrac{\text{distance travelled by the compound}}{\text{distance travelled by the solvent front}}$
retention time	in chromatography, the time each component remains in the column
secondary structure	of a protein, relates to the orderly, hydrogen-bonded arrangements between peptide chains resulting in either a helix or a pleated sheet
shielded	in n.m.r., a nucleus is said to be shielded when the electron density surrounding it is increased, giving rise to an upfield shift (smaller δ value)
singlet	in n.m.r., a peak that is not split
solvent front	in paper chromatography, the position reached by the leading edge of the solvent after separation has occurred
spin–spin coupling	in n.m.r., the interaction between the nuclear spins of non-equivalent hydrogen atoms on adjacent carbon atoms
splitting	in n.m.r., the splitting of an absorption signal (a peak) into more complex patterns as a result of coupling between neighbouring nuclear spins
stationary phase	in chromatography, the fixed phase through which passes the moving or mobile phase
step-growth polymer	see *condensation polymer*
stereoisomerism	occurs when molecules with the same structural formula have bonds arranged differently in space (see *E–Z stereoisomerism* and *optical isomerism*)
stereoisomers	are compounds which have the same structural formula but have bonds arranged differently in space
stoichiometric coefficient	the number moles of a species as shown in a balanced chemical equation
stoichiometric point	see *equivalence point*
strong acid/base	one that is (almost) completely dissociated in aqueous solution
structural isomerism	occurs when the component atoms are arranged differently in molecules having the same molecular formula
structural isomers	compounds with the same molecular formula but different structures
surfactant	a wetting agent, containing hydrophobic and hydrophilic groups, able to lower the surface tension of a liquid and the interfacial tension between two liquids; the name is derived from *surface acting agent*
Terylene	a polyester, used in permanent-press fabrics, derived from benzene-1,4-dicarboxylic (*terephthalic*) acid and ethane-1,2-diol
thin-layer chromatography (TLC)	involves a thin layer of a polar, adsorbent material coated on to a glass plate or on to an aluminium or plastic sheet (the stationary phase) and a solvent (the moving phase)
titration curve	a plot of the pH of an acid/base against the volume of base/acid added
transesterification	a reversible reaction in which an ester reacts with an alcohol, usually in excess, to form a new ester and a new alcohol

et	in n.m.r., a peak that is split into three parts
unimolecular step	a first-order step in a reaction mechanism
vinyl	the old name for ethenyl ($-CH=CH_2$)
weak acid approximation	when K_a is small, $K_a \approx \dfrac{[H^+]^2}{[HA]_{tot}}$
weak acid/base	one that is only partially dissociated in aqueous solution
zwitterion	a dipolar ion that has both a positive and a negative charge, especially an amino acid in neutral solution